TEST INSTRUMENTS
APPLICATIONS MANUAL

AMERICAN TECHNICAL PUBLISHERS, INC.
HOMEWOOD, ILLINOIS 60430-4600

Glen A. Mazur

© 2006 by American Technical Publishers, Inc. and the National Joint Apprenticeship & Training Committee for the Electrical Industry
All rights reserved

1 2 3 4 5 6 7 8 9 – 06 – 9

Printed in the United States of America

ISBN 978-0-8269-1326-5

 This book is printed on 10% recycled paper.

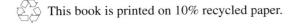

Test Instruments Applications Manual

Contents

1 Introduction to Test Instruments — 1

2 General Use Test Instruments — 41

3 Voice-Data-Video (VDV) Test Instruments — 67

Test Instruments Applications Manual includes application information and hands-on activities that expand on content presented in *Test Instruments*. *Test Instruments* provides extensive information on each of the test instruments that is necessary for completing the applications, including important safety considerations and procedures for use in the field.

The Applications Manual consists of review questions and applications. The review questions test the comprehension of important concepts from the textbook that form the basis for the applications. Types of review questions include short answer, true-false, and long answer (requiring one to three complete sentences).

The applications focus on the use of test instruments typically found in industry. These applications build on the textbook by addressing technical and regulatory information, and procedures for performing calculations and troubleshooting. The Manual introduces situations in which test instruments must be used and stresses testing procedures, reading prediction, and troubleshooting analysis. Students may be required to choose the most appropriate test instrument, connect the test instrument, select the correct testing function, and predict and analyze the readings.

The Appendix includes useful reference data and enlarged illustrations of selected commonly used test instruments. Answers to all review and activity questions are given in *Test Instruments Applications Manual Answer Key*.

Some applications describe using test instruments to ensure that electrical installations conform to *National Electrical Code® (NEC®)* standards. The purpose of the *NEC®* is to protect people and property from hazards that arise from the use of electricity. *National Electrical Code®* and *NEC®* are registered trademarks of the National Fire Protection Association, Inc., Quincy, MA 02169.

Test Instruments and *Test Instruments Application Manual* are two of many high-quality training products available from American Technical Publishers, Inc. To obtain information about related training products, visit the American Tech web site at www.go2atp.com.

The Publisher

Review Questions

Introduction to Test Instruments 1

Name_____ Date_____

_____ **1.** What is the abbreviation for milliamperes?

T F **2.** Digital meters are more popular than analog meters because digital meters typically have greater measuring capabilities.

T F **3.** The caution signal word indicates an imminently hazardous situation that, if not avoided, results in death or serious injury.

_____ **4.** What is the symbol for ohms?

_____ **5.** Which type of meter, analog or digital, is more likely to be affected by electrical noise?

T F **6.** Tagout physically prevents energy sources from being energized.

_____ **7.** What unit is used to measure capacitance?

_____ **8.** Which type of meter, analog or digital, is more likely to warn the user if the function switch is improperly set?

_____ **9.** What is the maximum allowable resistance for test leads?

_____ **10.** What unit is used to measure frequency?

T F **11.** A high CAT rating indicates that a test instrument can be used in a circuit with high power, great possibility of short circuits, and high energy transients.

_____ **12.** Which unit prefix indicates one million times the base unit?

T F **13.** Test instruments should be periodically tested for accuracy.

_____ **14.** Which unit prefix indicates one thousandth of the base unit?

_____ **15.** What type of additional personal protective equipment must be worn when working around electrical systems with a risk of arc flash?

_____ **16.** What electrical unit is used to rate rubber insulating gloves?

T F **17.** Leather protector gloves are used to protect wearers from electrical shock.

_____ **18.** Which organization publishes the National Electrical Code®?

_____ **19.** Which two temperature scales are most commonly used by test instruments?

T F **20.** Both the function switch and the connection of the test leads must match the desired measurement before testing a circuit.

21. Why do some meters include fuses?

22. Why is it often more convenient to use unit prefixes to express measurements?

23. What are ghost voltages?

24. What is grounding?

25. What is the difference between lockout and tagout?

Applications

Introduction to Test Instruments 1

Name_____   Date_____

Application 1-1   *Electrical Prefixes*

Prefixes are used with electrical and electronic components and test instruments to simplify large and small numbers. The prefixes most commonly used in the electrical field are mega- (M), kilo- (k), milli- (m), and micro- (μ). Some test instruments also use other, less common prefixes for certain measurements, such as pico- (p) for capacitor measurements and giga- (G) for high resistance measurements. For example, one picofarad (1 pF) is equal to 0.000000000001 F and one gigohm (1 GΩ) is equal to 1,000,000,000 Ω. See Metric Prefixes.

METRIC PREFIXES			
Multiples and Submultiples	**Prefixes**	**Symbols**	**Meaning**
$1,000,000,000,000 = 10^{12}$	tera	T	trillion
$1,000,000,000 = 10^{9}$	giga	G	billion
$1,000,000 = 10^{6}$	mega	M	million
$1000 = 10^{3}$	kilo	k	thousand
$100 = 10^{2}$	hecto	h	hundred
$10 = 10^{1}$	deka	da	ten
Base Unit $1 = 10^{0}$	—	—	—
$.1 = 10^{-1}$	deci	d	tenth
$.01 = 10^{-2}$	centi	c	hundredth
$.001 = 10^{-3}$	milli	m	thousandth
$.000001 = 10^{-6}$	micro	μ	millionth
$.000000001 = 10^{-9}$	nano	n	billionth
$.000000000001 = 10^{-12}$	pico	p	trillionth

Very small or very large numerical values can be awkward to use or convey, but can be made more manageable by changing the measuring unit. A base unit is any unit that does not include a prefix, such as 0.015 A, 1500 V, and 2,300,000 W. Values can be simplified by converting the numerical value and adding a prefix to the unit symbol. For example, 0.015 A = 15 mA, 1500 V = 1.5 kV, and 2,300,000 W = 2.3 MW. See Metric Conversions.

METRIC CONVERSIONS											

Initial Units	Final Units										
	giga	mega	kilo	hecto	deka	base unit	deci	centi	milli	micro	nano	pico
giga		3R	6R	7R	8R	9R	10R	11R	12R	15R	18R	21R
mega	3L		3R	4R	5R	6R	7R	8R	9R	12R	15R	18R
kilo	6L	3L		1R	2R	3R	4R	5R	6R	9R	12R	15R
hecto	7L	4L	1L		1R	2R	3R	4R	5R	8R	11R	14R
deka	8L	5L	2L	1L		1R	2R	3R	4R	7R	10R	13R
base unit	9L	6L	3L	2L	1L		1R	2R	3R	6R	9R	12R
deci	10L	7L	4L	3L	2L	1L		1R	2R	5R	8R	11R
centi	11L	8L	5L	4L	3L	2L	1L		1R	4R	7R	10R
milli	12L	9L	6L	5L	4L	3L	2L	1L		3R	6R	9R
micro	15L	12L	9L	8L	7L	6L	5L	4L	3L		3R	6R
nano	18L	15L	12L	11L	10L	9L	8L	7L	6L	3L		3R
pico	21L	18L	15L	14L	13L	12L	11L	10L	9L	6L	3L	

A conversion table can be used to quickly change between a base unit and equivalent units that use prefixes. For each prefix combination, the table lists the direction and number of places the decimal point is moved to change the numerical value. To convert a large numerical value into a smaller number with large units, the decimal point is moved to the left and the corresponding prefix (k, M, G, etc.) is added. To convert a small numerical value into a larger number with small units, the decimal point is moved to the right and the corresponding prefix (m, µ, p, etc.) is added.

For example, to change 470,000 W to kW, the decimal point is moved three places to the left and the prefix "k" added (470,000 W = 470 kW). Likewise, to change 0.024 A to mA, the decimal point is moved three places to the right and the prefix "m" added (0.024 A = 24 mA).

Convert each of the measurements.

1. _____ 0.420 mA = ___ A

2. _____ 0.420 mA = ___ µA

3. _____ 1.230 kΩ = ___ Ω

4. _____ 1.230 kΩ = ___ MΩ

5. _____ 0.954 MHz = ___ kHz

6. _____ 0.954 MHz = ___ Hz

7. _____ 351 ms = ___ s

8. _____ 351 ms = ___ µs

9. _____ 45 µF = ___ F

10. _____ 45 µF = ___ pF

11. _____ 0.640 V = ___ mV

12. _____ 0.640 V = ___ µV

13. _____ 21 µA = ___ A

14. _____ 21 µA = ___ mA

15. _____ 572 MΩ = ___ GΩ

16. _____ 572 MΩ = ___ kΩ

Electrical Quantities

Test instruments are used to measure electrical quantities, such as the voltage in a circuit. When voltage is a variable in electrical formulas, such as Ohm's law ($E = I \times R$) or the power formula ($P = E \times I$), it is represented by the capital letter "E," for electromotive force. However, when using a test instrument to measure voltage, the unit symbol for voltage (V) is used on the meter and meter display, not the letter "E."

For each electrical quantity being measured by a test instrument, it is important to know the unit of measurement and the abbreviations used to represent the electrical quantity on both the test instrument and in an electrical formula. See Common Electrical Quantities.

COMMON ELECTRICAL QUANTITIES		
Variable	**Name**	**Unit of Measure and Abbreviation**
E	voltage	volt — V
I	current	ampere — A
R	resistance	ohm — Ω
P	power	watt — W
P	power (apparent)	volt-amp — VA
C	capacitance	farad — F
L	inductance	henry — H
Z	impedance	ohm — Ω
G	conductance	siemens — S
f	frequency	hertz — Hz
T	period	second — s

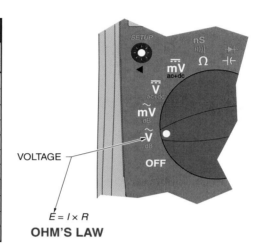

VOLTAGE

$E = I \times R$

OHM'S LAW

For each of the symbols on the test instruments, state the name of the electrical quantity being measured, the unit of measurement for the electrical quantity being measured, and the equivalent abbreviation for the electrical quantity being measured that is used as a variable in electrical formulas. For example, a meter function switch set to "V" will measure "voltage" in units of "volts," which is represented by the abbreviation "E."

1. _____   What is the name of the electrical quantity the test instrument is set to measure?

2. _____   What is the base unit of measurement for the electrical quantity to be measured?

3. _____   What is the abbreviation for the electrical quantity being measured, as it is used in an electrical formula (as a variable)?

4. _____ What is the name of the electrical quantity the test instrument is set to measure?

5. _____ What is the base unit of measurement for the electrical quantity to be measured?

6. _____ What is the abbreviation for the electrical quantity being measured, as it is used in an electrical formula (as a variable)?

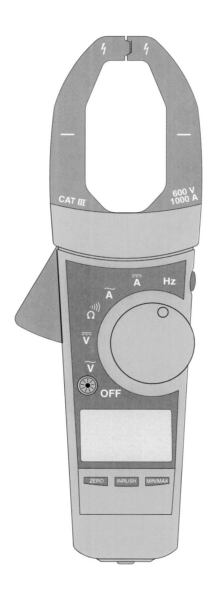

_____ **7.** What is the name of the electrical quantity the test instrument is set to measure?

_____ **8.** What is the base unit of measurement for the electrical quantity to be measured?

_____ **9.** What is the abbreviation for the electrical quantity being measured, as it is used in an electrical formula (as a variable)?

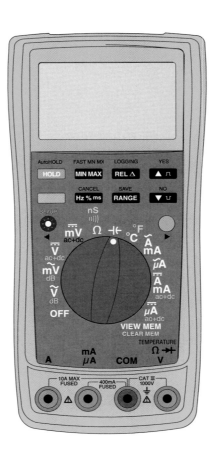

Application 1-3 *Test Instrument Settings and Connections*

Test instruments can take safe and accurate measurements only if they are properly set to the electrical measurement to be measured, the meter test leads are connected to the correct meter jacks, and the meter is properly connected to the circuit to be measured. One of the most common mistakes made when using a test instrument is the meter's function switch setting not matching the connection of the meter's test leads—for example, setting a meter to measure voltage (V) but connecting the test leads to the current (A) jacks. Another common mistake is setting the meter to measure VAC and connecting it to a VDC part of a circuit (or setting the meter to VDC and connecting it to a VAC part of a circuit).

1. Set (draw in) the function switch on the multimeter to use the continuity test mode to test the normally open (NO) contacts on a pushbutton before it is wired into a circuit.

2. Connect the two test leads to the correct jacks on the multimeter.

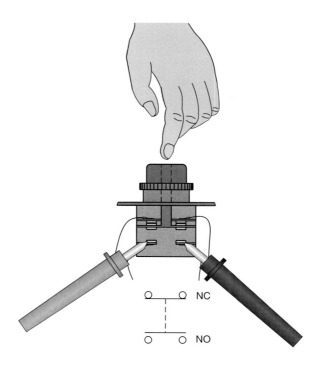

3. Set (draw in) the function switch on the multimeter to take voltage measurements on the step-down transformer in a door chime circuit.

4. Connect the two test leads to the correct jacks on the multimeter.

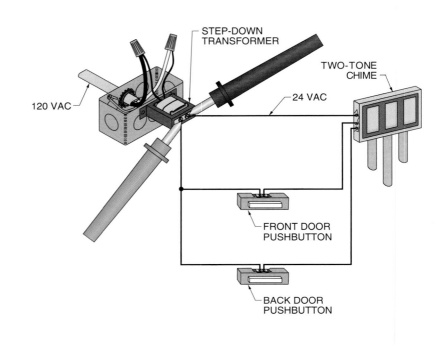

5. Set (draw in) the function switch on Multimeter 1 to take in-line current measurements to determine the current draw of the motor starter coil.

6. Connect the two test leads connected to the motor starter coil circuit to the correct jacks on Multimeter 1.

7. Set (draw in) the function switch on Multimeter 2 to take voltage measurements on the circuit connecting the solid-state relay to the PLC computer.

8. Connect the two test leads connected to the SSR to the correct jacks on Multimeter 2.

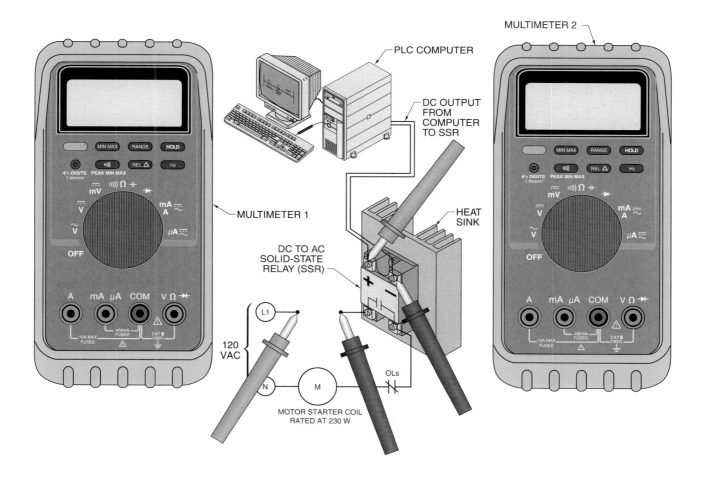

Per NFPA 70E, personal protective and safety requirements must be followed when working around energized electrical circuits. Approved rubber electrical gloves must be worn anytime electrical measurements are taken on energized circuits in which there is a chance of an electrical shock (usually 50 V and above).

The three types of gloves an electrician may wear include insulated (rubber) gloves, cotton liners, and leather outer gloves. An insulated (rubber) glove (required) provides a high enough resistance to prevent electricity from entering the hand. A cotton liner (optional) is worn inside the insulated glove to add comfort and aid hand movement. A leather outer glove must be worn to protect the insulated rubber glove.

An individual's glove size must be known in order to select a glove that fits best and provides optimum comfort and protection. To determine glove size, the hand is held flat with the fingers together and thumb extended. See Glove Sizing. The circumference around the knuckles is measured and rounded up to the nearest ½″ (7, 7½, 8, etc.). The glove size is obtained by adding ½″ to the rounded measurement. See Electrical Glove Sizes. A string may be used if a flexible tape measure is not available. The length of the string is measured with a standard tape measure or ruler.

Dielectric rubber gloves are typically seamless with a curved hand design for ease of use with hot and live electrical equipment. Gloves are available in solid black or in two-tone options (yellow/black or orange/black) that show color when the glove is worn through or damaged. Gloves are tested in accordance with ASTM D120 specifications.

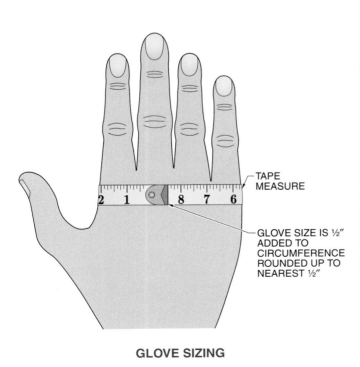

GLOVE SIZING

ELECTRICAL GLOVE SIZES		
Model Number	**Size**	**Protection***
7-1000	7	1000
7-20k	7	20,000
7.5-1000	7.5	1000
7.5-20k	7.5	20,000
8-1000	8	1000
8-20k	8	20,000
8.5-1000	8.5	1000
8.5-20k	8.5	20,000
9-1000	9	1000
9-20k	9	20,000
9.5-1000	9.5	1000
9.5-20k	9.5	20,000
10-1000	10	1000
10-20k	10	20,000
10.5-1000	10.5	1000
10.5-20k	10.5	20,000
11-1000	11	1000
11-20k	11	20,000
11.5-1000	11.5	1000
11.5-20k	11.5	20,000
12-1000	12	1000
12-20k	12	20,000

* in V

Determine the proper glove size and model number.

1. _____ What is the proper glove size for a hand with a circumference of 6¼″?

2. _____ Which glove model is required for a hand with a circumference of 6¼″ when working around 120 V or less?

3. _____ What is the proper glove size for a hand with a circumference of 10⅓″?

4. _____ Which glove model is required for a hand with a circumference of 10⅓″ when working around 480 V or less?

5. _____ What is the proper glove size for a hand with a circumference of 8¾″?

6. _____ Which glove model is required for a hand with a circumference of 8¾″ when working around 1200 V or less?

7. _____ What is the proper glove size for a hand with a circumference of 9⅛″?

8. _____ Which glove model is required for a hand with a circumference of 9⅛″ when working around 1.5 kV or less?

Electrical circuits are used to transfer, switch, control, or convert electrical energy into some other form of energy, such as light, sound, or mechanical motion. These processes also produce heat. Electric heat lamps and electric heating elements produce heat intentionally. Unwanted heat is produced in an electrical circuit any time current flowing through a circuit encounters resistance, such as when current flows through conductors (wire) and encounters a bad (loose, corroded, etc.) connection or a faulty switch (having worn contacts). The higher the resistance of the undersized conductor, bad connection, or faulty switch, the greater the amount of heat produced at that point.

Test instruments can be used to measure the amount of heat at different locations within an electrical system. By taking temperature measurements throughout an electrical circuit, existing problems can be found (troubleshooting), and potential future problems can be found (preventive maintenance).

Temperature is usually measured in degrees Fahrenheit (°F) or degrees Celsius (°C). Some temperature measurement instruments can be set to display a temperature measurement in both degrees Fahrenheit and degrees Celsius, and others can only display the temperature measurement in either degrees Fahrenheit or degrees Celsius, but not both. It is important to know how to convert between degrees Fahrenheit and degrees Celsius when taking and recording temperature measurements. See Temperature Conversion.

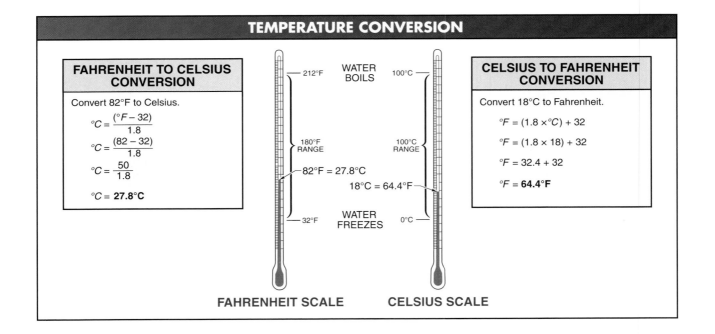

TEMPERATURE CONVERSION

FAHRENHEIT TO CELSIUS CONVERSION

Convert 82°F to Celsius.

$$°C = \frac{(°F - 32)}{1.8}$$

$$°C = \frac{(82 - 32)}{1.8}$$

$$°C = \frac{50}{1.8}$$

$$°C = \textbf{27.8°C}$$

CELSIUS TO FAHRENHEIT CONVERSION

Convert 18°C to Fahrenheit.

$$°F = (1.8 × °C) + 32$$

$$°F = (1.8 × 18) + 32$$

$$°F = 32.4 + 32$$

$$°F = \textbf{64.4°F}$$

212°F — WATER BOILS — 100°C

180°F RANGE — 100°C RANGE

82°F = 27.8°C

18°C = 64.4°F

32°F — WATER FREEZES — 0°C

FAHRENHEIT SCALE   CELSIUS SCALE

Convert each of the temperature measurements.

1. _____ What is the electrical connection temperature in degrees Celsius?

2. _____ What is the ambient temperature in degrees Celsius?

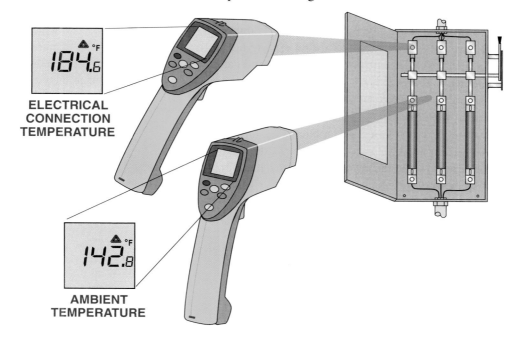

3. _____ What is the electrical connection temperature in degrees Fahrenheit?

4. _____ What is the ambient temperature in degrees Fahrenheit?

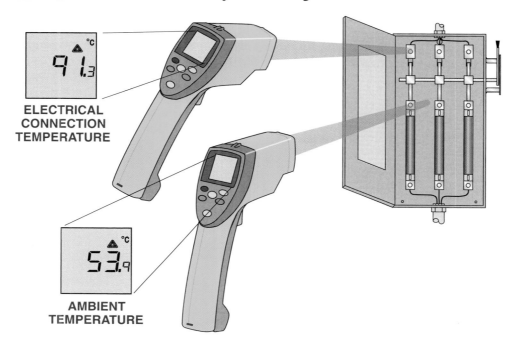

Application 1-6   *MIN MAX Recording Time*

Test instruments take instantaneous measurements. For example, a voltmeter connected to a receptacle (outlet) will display the voltage at the receptacle at that moment. Instantaneous measurements are useful for most situations, but sometimes measurements over a period of time are required to diagnose some problems. For example, the voltage at receptacles will fluctuate within an acceptable level (+5% to –10%), even if the receptacle is properly wired and operating normally. However, if there is a power quality or wiring problem, large voltage variations can occur and cause problems in loads connected to the receptacle, such as causing computers to reboot. In order to find a voltage variation problem, measurements must be taken over a period of time.

Any multimeter with a MIN MAX recording function can be used to take voltage (or other electrical quantity) measurements over time. The meter saves any new minimum or maximum values that last for at least a certain period of time. This measurement duration depends on the meter's make and model and is specified in milliseconds on the meter or in the manual. Shorter durations indicate that the meter is capable of detecting shorter fluctuations.

Peak voltage measurements are taken over time to detect and measure transients on a power line. A transient voltage (voltage spike) is a temporary, undesirable voltage in an electrical circuit. Transient voltages range from a few volts to several thousand volts and last from a few microseconds up to a few milliseconds. In order to capture and record transients (peak voltages), the meter must have a much faster minimum measurement duration than is required for taking general MIN MAX recordings. Meters with PEAK MIN MAX functions have the capability of changing the duration to a smaller value for measuring transients.

1. _____   How long (in sec) must a new maximum or minimum voltage last before this multimeter records a new measured value?

2. _____   Could the MIN MAX function on this multimeter record a momentary power interruption on a standard 60 Hz power line that lasted for three complete cycles?

3. _____   For this multimeter, for how many complete cycles would a momentary power interruption have to exist before the MIN MAX function recorded the voltage change?

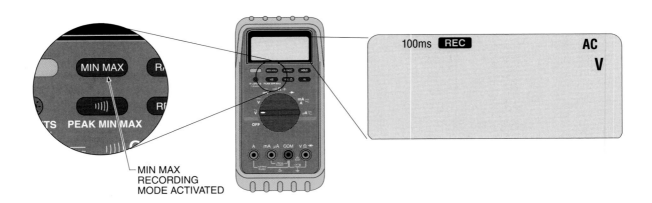

4. _____ How long (in sec) must a new maximum or minimum voltage last before this multimeter records a new measured value?

5. _____ If there are 60 complete cycles in one second (60 Hz), for how many milliseconds does one cycle (1 Hz) last?

6. _____ Could the PEAK MIN MAX function on this multimeter record a momentary power interruption on a standard 60 Hz power line that lasted for one-half cycle?

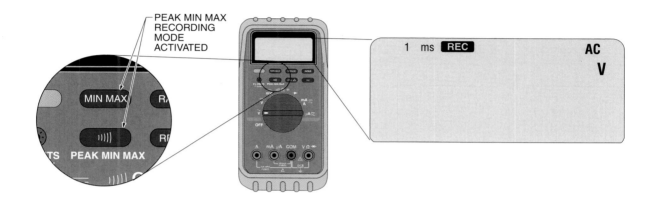

IEC Standard 1010 classifies the applications in which test instruments may be used into four overvoltage installation categories (CAT I–CAT IV). They categorize the magnitude of transient voltage a test instrument must withstand when used on a power distribution system. Test instruments are designed and marked for the maximum category in which they can safely be used. Applications require test instruments with a CAT rating the same or higher than the application category. See IEC 1010 Overvoltage Installation Categories.

IEC 1010 OVERVOLTAGE INSTALLATION CATEGORIES

Category	In Brief	Examples
CAT I	Electronic	• Protected electronic equipment • Equipment connected to (source) circuits in which measures are taken to limit transient overvoltage to an appropriately low level • Any high-voltage, low-energy source derived from a high-winding-resistance transformer, such as the high-voltage section of a copier
CAT II	1φ receptacle-connected loads	• Appliances, portable tools, and other household and similar loads • Outlets and long branch circuits • Outlets at more than 30′ (10 m) from CAT III source • Outlets at more than 60′ (20 m) from CAT IV source
CAT III	3φ distribution, including 1φ commercial lighting	• Equipment in fixed installations, such as switchgear and polyphase motors • Bus and feeder in industrial plants • Feeders and short branch circuits and distribution panel devices • Lighting systems in larger buildings • Appliance outlets with short connections to service entrance
CAT IV	3φ at utility connection, any outdoor conductors	• Refers to the origin of installation, where low-voltage connection is made to utility power • Electric meters, primary overcurrent protection equipment • Outside and service entrance, service drop from pole to building, run between meter and panel • Overhead line to detached building

1. _____ What is the minimum CAT rating required for a test instrument taking measurements at an electric meter to test for proper voltage on service entrance conductors?

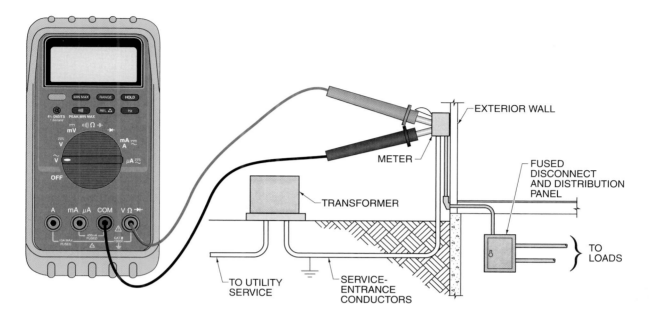

2. _____ What is the minimum CAT rating required for a test instrument taking measurements on a soft drink beverage gun to test electrical connections?

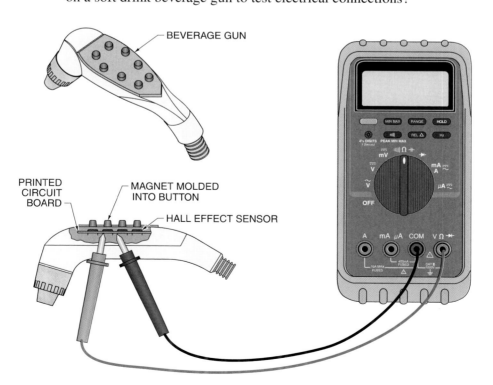

3. _____ What is the minimum CAT rating required for a test instrument taking measurements at a receptacle to make sure it is wired correctly?

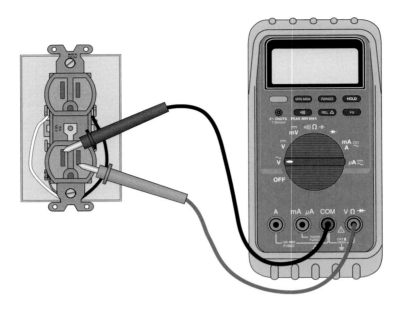

4. _____ What is the minimum CAT rating required for a test instrument taking measurements on photoelectric switches used to detect truck positions at a loading dock?

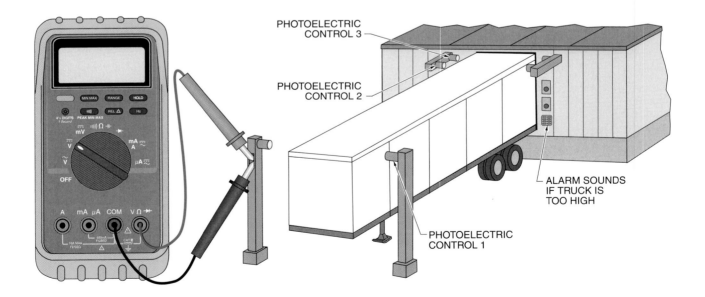

Application 1-8

Lockout is the process of removing the source of electrical power and installing a lock that prevents the power from being turned on. Tagout is the process of placing a danger tag on the source of power, which indicates that the power may not be turned on until the danger tag is removed. Lockout/tagout procedures help prevent an electrical shock when working on electrical equipment and are required whenever work can be done without power. However, lockouts and tagouts cannot be used when the work requires electrical power to be on in order to take electrical measurements with test instruments.

Determine if a lockout/tagout device is required and appropriate for each service call. Write "YES" if lockout/tagout is required and write "NO" if lockout/tagout is not required and cannot be used for the service call.

1. _____ Is lockout/tagout required for a service call where cartridge heating elements are changed to a larger size to produce more heat on the press platens?

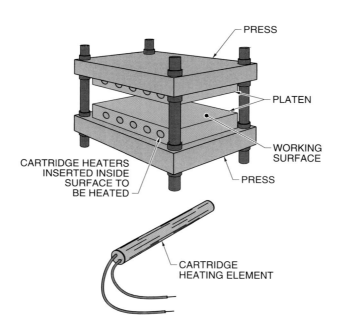

2. _____ Is lockout/tagout required for a service call where voltage is measured at the heating element to make sure the SCRs are properly controlling the voltage level at the heating element?

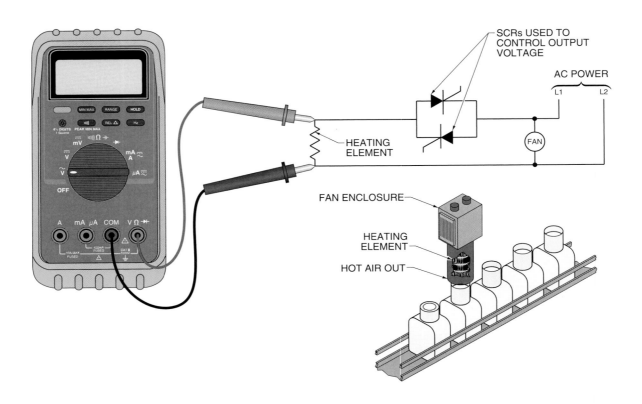

3. _____ Is lockout/tagout required for a service call where the condition (tightness and wear) of the motor/compressor belt is checked?

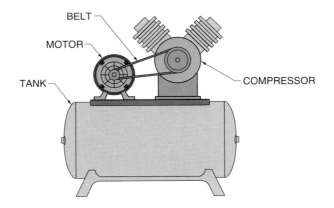

4. _____ Is lockout/tagout required for a service call where a clamp-on ammeter is used to measure the current draw of the conveyor motor at different product load conditions?

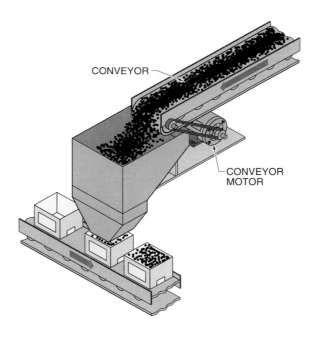

CONVEYOR

CONVEYOR
MOTOR

A conductor is a low-resistance material that carries electricity to different parts of a circuit. Copper is the most common conductor material used for inside wiring because it has a very low resistance. Low resistance means less heat and less voltage drop across the conductor.

Aluminum is also used for conductors, but must be sized larger than copper conductors to compensate for aluminum's higher material resistance. However, since it is much lighter than copper (even when sized up), aluminum is used for outside power distribution lines where weight is a major factor. Copper-clad aluminum conductors have copper bonded to aluminum cores, which balances the lower resistance of copper with the lower weight of aluminum.

Conductor Sizes

Conductors are sized based on the American Wire Gauge (AWG) numbering system. See Conductor Sizes. Smaller AWG numbers indicate larger conductors and more current-carrying capacity. For example, a No. 12 AWG copper conductor is larger in diameter than a No. 14 copper conductor and may carry more current. Conductors that are No. 8 and smaller may be either solid or stranded. Conductors that are larger than No. 8 are stranded. See Stranded and Solid Conductors.

CONDUCTOR SIZES					
AWG	Diameter*	Area	AWG	Diameter*	Area
00 (2/0)	0.3648	●	8	0.1285	●
0 (1/0)	0.3249	●	10	0.1019	●
1	0.2893	●	12	0.0808	●
2	0.2576	●	14	0.0641	●
3	0.2294	●	16	0.0508	●
4	0.2043	●	18	0.0403	●
6	0.1620	●	20	0.0320	●

* in in.

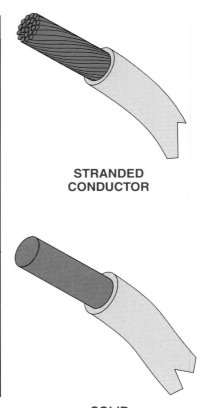

STRANDED CONDUCTOR

SOLID CONDUCTOR

Conductors that are No. 18 and No. 16 are normally used for the power cords of small appliances, plug-in lamps, and speakers. Conductors that are No. 14 and No. 12 are used for wiring most lighting circuits and supplying power to standard receptacle outlets. Conductors that are No. 10 through No. 4 are used for wiring residential and commercial appliances, such as electric ranges, water heaters, furnaces, and air conditioners. They are also used for supplying power to subpanels and large motors. Conductors that are No. 3 and larger are used for supplying power to main service panels.

Ampacity

Ampacity is a rating based on the amount of continuous current that insulated conductors can safely carry without damaging the insulation. Ampacity depends on the conductor material and size, the type of insulation, the ambient temperature at the conductor location, and the number of current-carrying conductors run together. NEC® tables list the ampacities of various conductors and can be used to determine the minimum conductor size or type for a particular installation. See Ampacities of Insulated Conductors.

AMPACITIES OF INSULATED CONDUCTORS*

Type of Insulation	Types TW, UF	Types RHW, THHW, THW, THWN, XHHW, USE, ZW	Types TBS, SA, SIS, FEP, FEPB, MI, RHH, RHW-2, THHN, THW-2, THWN-2, USE-2, XHH, XHHW-2, ZW-2	Types TW, UF	Types RHW, THHW, THW, THWN, XHHW, USE	Types TBS, SA, SIS, THHN, THW-2, THWN-2, RHH, RHW-2, USE-2, XHH, XHHW-2, ZW-2
AWG	**COPPER**			**ALUMINUM OR COPPER-CLAD ALUMINUM**		
18	—	—	14	—	—	—
16	—	—	18	—	—	—
14	20	20	25	—	—	—
12	25	25	30	20	20	25
10	30	35	40	25	30	35
8	40	50	55	30	40	45
6	55	65	75	40	50	60
4	70	85	95	55	65	75
3	85	100	110	65	75	85
2	95	115	130	75	90	100
1	110	130	150	85	100	115
1/0	125	150	170	100	120	135
2/0	145	175	195	115	135	150
3/0	165	200	225	130	155	175
4/0	195	230	260	150	180	205
Ambient Temperature†	**Correction Factor for Ambient Temperature**					
21–25	1.08	1.05	1.04	1.08	1.05	1.04
26–30	1.00	1.00	1.00	1.00	1.00	1.00
31–35	0.91	0.94	0.96	0.91	0.94	0.96
36–40	0.82	0.88	0.91	0.82	0.88	0.91
41–45	0.71	0.82	0.87	0.71	0.82	0.87
46–50	0.58	0.75	0.82	0.58	0.75	0.82
51–55	0.41	0.67	0.76	0.41	0.67	0.76
56–60	—	0.58	0.71	—	0.58	0.71
61–70	—	0.33	0.58	—	0.33	0.58
71–80	—	—	0.41	—	—	0.41

* Based on ambient temperature of 30°C (86°F) and not more than three current-carrying conductors in a raceway, cable, or earth (directly buried).
† in °C

Conductors that carry more current create more heat, requiring insulation material that can withstand higher temperatures. Various insulation materials have different temperature breakdown ratings. The most common types of insulation material are designated TW, RHW, and SA.

Correction Factors

Additional conditions must be considered when determining conductor ampacity. If the ambient temperature at the installation location is higher than 30°C (86°F), the ampacity must be derated using a correction factor. The ambient temperature used to determine the correction factor should be the highest expected temperature for that location, which may be high near sources of heat, or for warm climates during summer months.

For example, an installation calls for a copper conductor with TW insulation that must carry 40 A of current. Using the table, the size of the conductor must be at least No. 8 AWG. However, the conductor will be installed near a bakery oven and the ambient temperature may reach as high as 120°F (49°C). The correction factor for this type of conductor at this temperature is 0.58. Therefore, the No. 8 conductor is derated to 23.2 A (40 A × 0.58 = 23.2 A). For a similar conductor to carry 40 A in this location, it must be a larger size. The minimum size for this installation is found by derating the ampacities of other conductors to find one with at least 40 A of corrected ampacity. In this case, at least a No. 4 TW conductor is required (70 A × 0.58 = 40.6 A). Another option is to use a conductor with insulation rated for higher temperatures, such as SA. A No. 8 copper conductor with SA insulation could be used instead (55 A × 0.82 = 45.1 A).

Another condition that must be accounted for is the number of current-carrying conductors in a raceway, cable, or underground because the heat created by conductors that are close together reduces their ampacity. When four or more current-carrying conductors are run together, a correction factor is applied to derate their ampacities. See Correction Factor for More than Three Current-Carrying Conductors. This correction is applied after ampacity is adjusted for ambient temperature. Non-current-carrying conductors, such as grounding conductors, are not counted when determining this correction factor.

CORRECTION FACTOR FOR MORE THAN THREE CURRENT-CARRYING CONDUCTORS	
Number of Current-Carrying Conductors	**Correction Factor**
4–6	0.80
7–9	0.70
10–20	0.50
21–30	0.45
31–40	0.40
Over 40	0.35

Reprinted with permission from NFPA 70-2005, the National Electrical Code®
Copyright© 2004, National Fire Protection Association, Quincy, MA 02169.
This reprinted material is not the official position of the NFPA on the referenced
subject which is represented solely by the standard in its entirety.

For example, if four copper TW current-carrying conductors are called for in the bakery installation, a correction factor of 0.80 must also be applied to the ampacity rating. The No. 4 conductor is now derated to 32.5 A, so it is no longer adequate. Therefore, No. 2 is the minimum size for this type of conductor for this installation (95 A × 0.58 = 55.1 A; 55.1 A × 0.80 = 44.1 A).

In each of the following conductor installation applications, the ambient temperature is measured during the highest expected normal high temperature time. Using the tables and the information provided, determine the minimum AWG conductor size for the installation application.

As with many electrical installations, the current value is not always directly given. Ohm's law and the power formula can be used to find an unknown electrical quantity when two other electrical quantities are given or measured.

1. _____ What is the minimum conductor size for this installation?

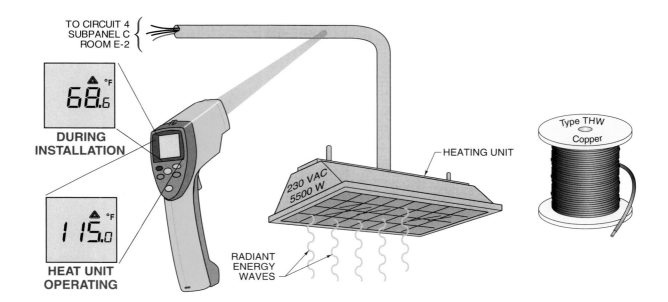

2. _____ What is the minimum conductor size for the branch circuit in this installation?

3. _____ What is the minimum conductor size for the power supply in this installation?

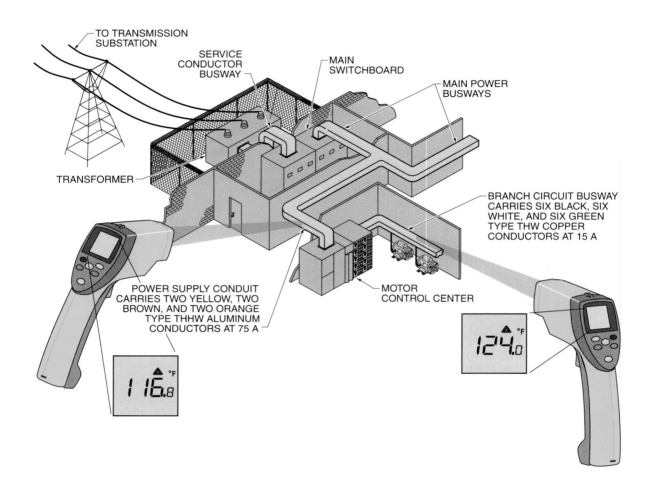

TO TRANSMISSION SUBSTATION

SERVICE CONDUCTOR BUSWAY

MAIN SWITCHBOARD

MAIN POWER BUSWAYS

TRANSFORMER

BRANCH CIRCUIT BUSWAY CARRIES SIX BLACK, SIX WHITE, AND SIX GREEN TYPE THW COPPER CONDUCTORS AT 15 A

POWER SUPPLY CONDUIT CARRIES TWO YELLOW, TWO BROWN, AND TWO ORANGE TYPE THHW ALUMINUM CONDUCTORS AT 75 A

MOTOR CONTROL CENTER

4. _____ What is the minimum conductor size for this installation?

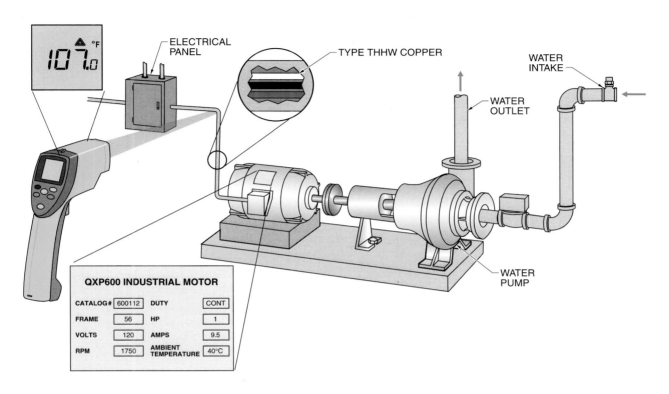

5. _____ What is the minimum conductor size for this installation?

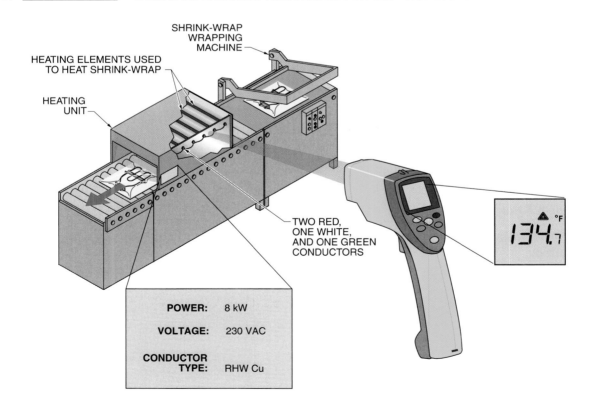

In addition to the ambient temperature and the number of current-carrying conductors, another factor limits the amount of current any given conductor can safely carry. The length of the conductor from the service panel (circuit breaker protecting the branch circuit) to the end of the run (furthest point) and back must be limited so that no more than 3% of the supplied voltage is dropped across the conductor. The amount of voltage drop depends on the current carried by the conductor and the resistance of the conductor. Resistances vary with conductor type and size. See Conductor Resistances.

CONDUCTOR RESISTANCES*				
AWG	**COPPER**		**ALUMINUM**	
	Solid	**Stranded**	**Solid**	**Stranded**
18	7.77	7.95	12.80	13.10
16	4.89	4.99	8.05	8.21
14	3.07	3.14	5.06	5.17
12	1.93	1.98	3.18	3.25
10	1.21	1.24	2.00	2.04
8	0.764	0.778	1.26	1.28
6	—	0.491	—	0.808
4	—	0.308	—	0.508
3	—	0.245	—	0.403
2	—	0.194	—	0.319
1	—	0.154	—	0.253

* in ohms per 1000 ft at 75°C (167°F)

Reprinted with permission from NFPA 70-2005, the National Electrical Code® Copyright© 2004, National Fire Protection Association, Quincy, MA 02169. This reprinted material is not the official position of the NFPA on the referenced subject which is represented solely by the standard in its entirety.

Calculating Voltage Drop

To determine the voltage drop across a conductor run, first calculate the total resistance of the conductor by applying the following formula:

$$R_R = R_C \times L$$

where

R_R = resistance of conductor for length of run (in Ω)

R_C = resistance of conductor per unit length (in Ω/ft)

L = length of run (in ft)

Then, using Ohm's law, the voltage drop across the conductor can be calculated by applying the following formula:

$$E_D = I \times R_R$$

where

E_D = voltage drop across conductor (in V)

I = current in conductor (in A)

R_R = resistance of conductor for length of run (in Ω)

For example, what is the voltage drop across a 300 ft run (150 ft each way) of No. 6 AWG stranded copper conductor carrying 55 A? From the table, a No. 6 copper conductor has a resistance of 0.491 Ω per 1000 ft at 75°F. This is equivalent to 0.000491 Ω/ft (0.491 Ω ÷ 1000 ft = 0.000491 Ω/ft).

$$R_R = R_C \times L$$
$$R_R = 0.000491 \text{ Ω/ft} \times 300 \text{ ft}$$
$$R_R = \mathbf{0.147 \text{ Ω}}$$

$$E_D = I \times R_R$$
$$E_D = 55 \text{ A} \times 0.147 \text{ Ω}$$
$$E_D = \mathbf{8.09 \text{ V}}$$

The percentage voltage drop can be calculated by applying the following formula:

$$\%E_D = \frac{E_D}{E} \times 100$$

where

$\%E_D$ = percentage voltage drop (in %)

E_D = voltage drop across conductor (in V)

E = voltage supplied at service panel (in V)

100 = conversion factor

For example, what is the percentage voltage drop of the same No. 6 copper conductor run with a supplied voltage of 230 V?

$$\%E_D = \frac{E_D}{E} \times 100$$

$$\%E_D = \frac{8.09 \text{ V}}{230 \text{ V}} \times 100$$

$$\%E_D = \mathbf{3.5\%}$$

A voltage drop of 3.5% is not acceptable under NEC® guidelines. To reduce the voltage drop, the conductor must be larger, the run shorter, the supplied voltage greater, or the current load less. Usually, the circuit voltage, current, and run length cannot be changed, so sizing up the conductor to reduce the resistance is the best choice. A No. 4 copper conductor in the same circuit would drop only 5.08 V (2.2%) and a No. 3 would drop 4.04 V (1.8%).

Measuring Voltage Drop

Calculating the voltage drop across a conductor to determine the correct conductor size should be done in the planning stage of an electrical project. However, conductor length is not always known accurately and the circuit may be modified later by adding or changing loads. Also, even if the conductor is properly sized for a circuit, any loose or corroded connections will increase the total resistance between the power panel and load. Therefore, the most accurate way to determine the voltage drop on a branch circuit is by taking voltage measurements.

Voltage is measured under no-load and full-load conditions to determine conductor voltage drop. To measure voltage drop in a branch circuit, apply the following procedure:

❶ Turn all loads connected to the branch circuit OFF.

❷ Measure the voltage at the furthest outlet (or load that is OFF) on the branch circuit to obtain the circuit's no-load voltage.

❸ Turn all loads connected to the branch circuit ON.

❹ Measure the voltage at the furthest outlet (or load that is ON) on the branch circuit to obtain the circuit's full-load voltage.

The percent of voltage drop in the branch circuit is calculated by applying the following formula:

$$\%V_D = \frac{V_{NL} - V_{FL}}{V_{NL}} \times 100$$

where

$\%V_D$ = percentage voltage drop (in V)

V_{NL} = no-load voltage (in V)

V_{FL} = full-load voltage (in V)

100 = conversion factor

For example, a branch circuit delivering power to a furnace has a measured full-load voltage of 112.5 V (furnace ON) and a measured no-load voltage of 115.8 V (furnace OFF). What is the percentage voltage drop?

$$\%V_D = \frac{V_{NL} - V_{FL}}{V_{NL}} \times 100$$

$$\%V_D = \frac{115.8 - 112.5}{115.8} \times 100$$

$$\%V_D = \frac{3.3}{115.8} \times 100$$

$$\%V_D = \mathbf{2.8\%}$$

The measured voltage drop across the conductor is less than 3%, so the conductor is properly sized for this circuit.

Determine the percent of voltage drop for each application from both the design specifications and using the measured values.

1. _____ What is the percent of voltage drop, as calculated from the design specifications?

2. _____ Is the calculated voltage drop within the acceptable limit?

3. _____ What is the percent of voltage drop, as determined from the measured values?

4. _____ Is the measured voltage drop within the acceptable limit?

5. What might account for the significant difference between the calculated and actual voltage drops?

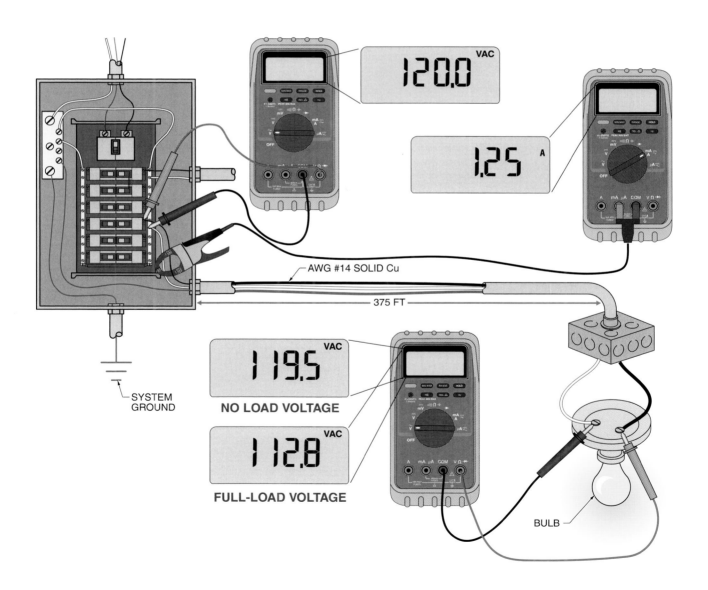

6. _____ What is the percent of voltage drop, as calculated from the design specifications?

7. _____ Is the calculated voltage drop within the acceptable limit?

8. _____ What is the percent of voltage drop, as determined from the measured values?

9. _____ Is the measured voltage drop within the acceptable limit?

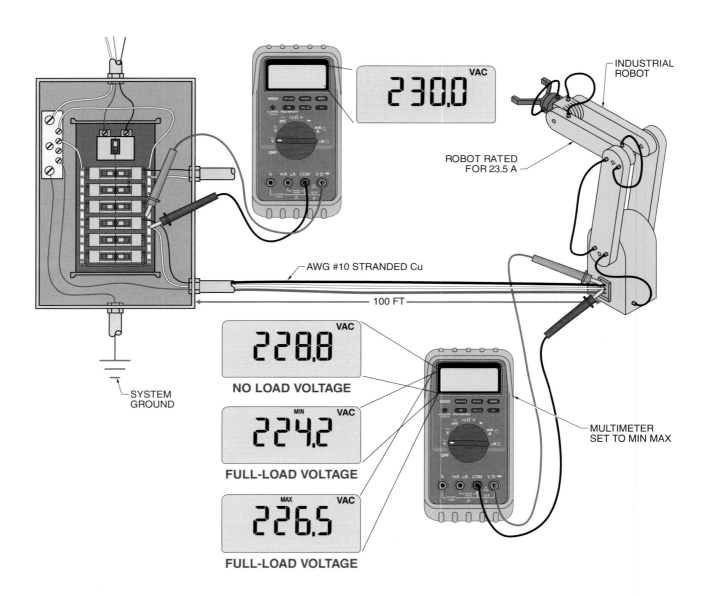

10. _____ What is the percent of voltage drop, as calculated from the design specifications?

11. _____ Is the calculated voltage drop within the acceptable limit?

12. _____ What is the percent of voltage drop, as determined from the measured values?

13. _____ Is the measured voltage drop within the acceptable limit?

14. What might account for the significant voltage drop in this circuit?

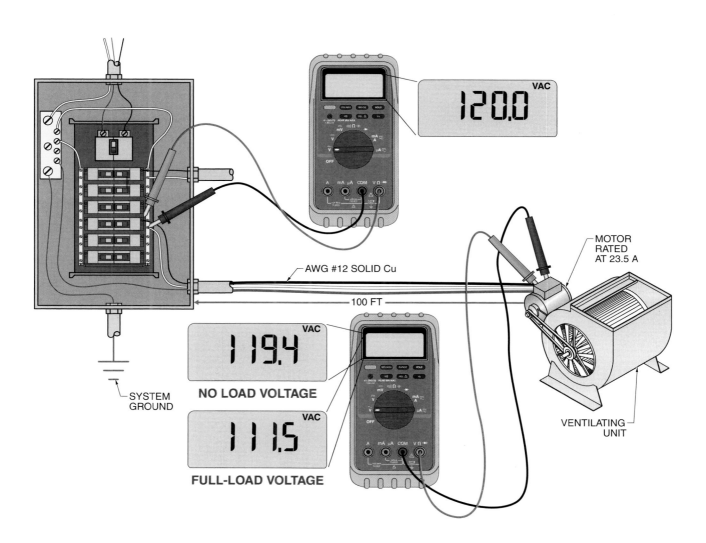

15. _____ What is the percent of voltage drop, as calculated from the design specifications?

16. _____ Is the calculated voltage drop within the acceptable limit?

17. _____ What is the percent of voltage drop, as determined from the measured values?

18. _____ Is the measured voltage drop within the acceptable limit?

19. _____ What would be the calculated percent of voltage drop if the conductors were AWG #12 stranded aluminum instead?

20. _____ Would the calculated voltage drop with the aluminum conductor be within the acceptable limit?

21. _____ If stranded aluminum wire were required, what size would be needed to keep the voltage drop below the limit?

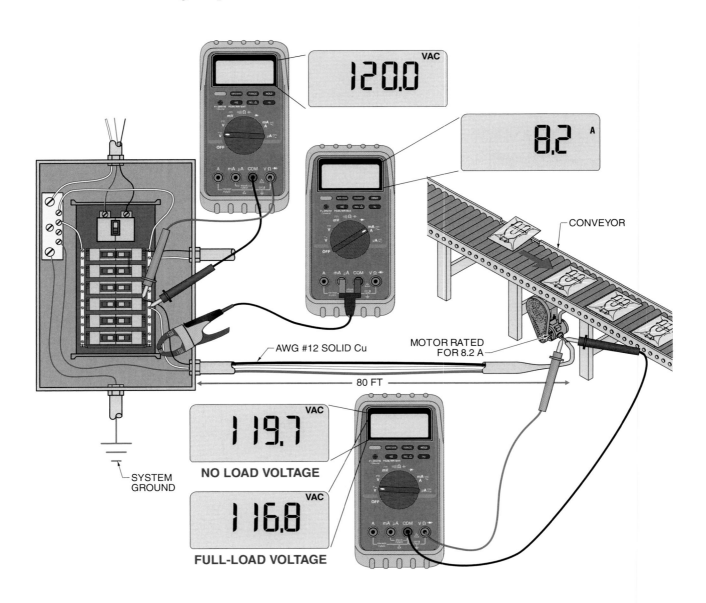

Fluorescent Lamp Output

The light output of fluorescent lamps is affected by both power quality and ambient temperature. A fluorescent lamp uses a transformer called a ballast to limit current flow and supply a high starting voltage for the lamp. A supply voltage higher or lower than the rating of the ballast affects the life of the lamp and ballast, and the lamp's light output (in lumens). The voltage supplied to a fluorescent lamp should be within ±7% of the lamp rating. See Fluorescent Lamp Fixture and Fluorescent Lamp Voltage Characteristics.

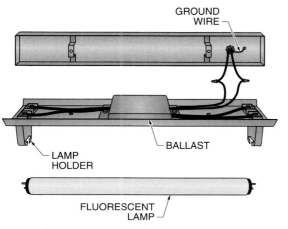

FLUORESCENT LAMP FIXTURE

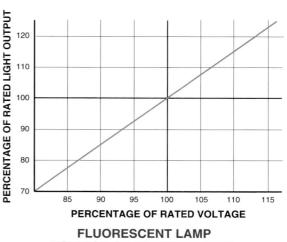

**FLUORESCENT LAMP
VOLTAGE CHARACTERISTICS**

Standard indoor fluorescent lamps are designed to deliver a peak light output at approximately 75°F ambient temperature, considered to be room temperature for most fluorescent lamp installations. Moderate changes in ambient temperature (50°F to 105°F) have little effect (less than 10%) on the light output. Temperatures lower than 50°F or higher than 105°F have a greater effect on light output. See Fluorescent Lamp Temperature Characteristics.

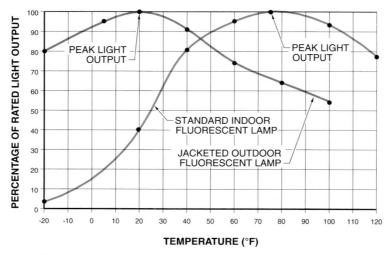

FLUORESCENT LAMP TEMPERATURE CHARACTERISTICS

Fluorescent lamps in cold conditions do not start well and put out less light. Cold lamps require more time to start because the cathode at the end of the lamp must heat up to release electrons into the mercury vapor. Then, even after the lamp is ON, it delivers less light because low temperature affects the mercury vapor pressure inside the lamp.

Outdoor-rated florescent lamps include an outer glass jacket. The jacket helps retain heat, which effectively shifts the peak light output to a lower ambient temperature point. Jacketed lamps are recommended for outdoor and cold indoor applications such as freezer warehouses, subways, and tunnels.

When accounting for both the ballast voltage and the ambient temperature to determine the actual light output, the two percentage values are multiplied. For example, if a low voltage reduces light output to 80% of the rated value and a cold environment reduces light output to 60% of the rated value, the effect of both together results in an actual light output of 48% of the rated value (80% × 60% = 48%).

Use the voltage measurement to determine the percentage of rated light output based on voltage and based on ambient temperature. Then determine the actual percentage of rated light output based on both voltage and temperature.

1. _____ What is the percentage of rated light output based on the ballast voltage?

2. _____ What is the percentage of rated light output based on the ambient temperature?

3. _____ What is the actual percentage of rated light output based on both the ballast voltage and ambient temperature?

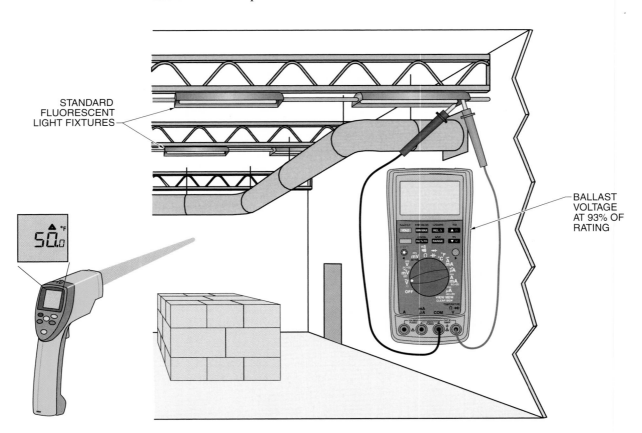

4. _____ What is the percentage of rated light output based on the ballast voltage?

5. _____ What is the percentage of rated light output based on the ambient temperature?

6. _____ What is the actual percentage of rated light output based on both the ballast voltage and ambient temperature?

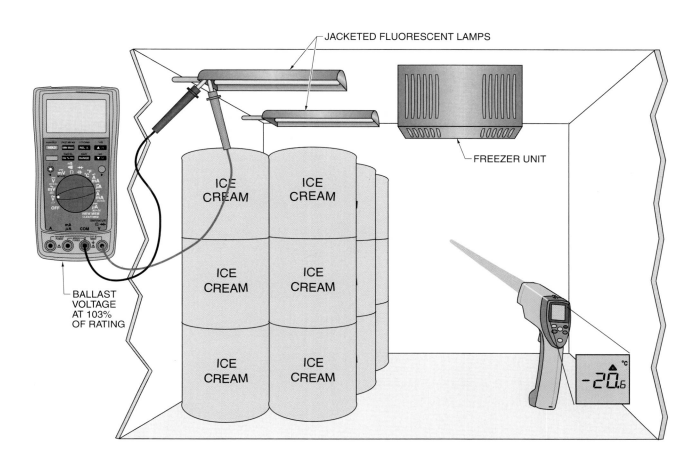

JACKETED FLUORESCENT LAMPS

FREEZER UNIT

ICE CREAM

BALLAST VOLTAGE AT 103% OF RATING

Name_____ Date_____

_____		**1.**	What type of voltage test instrument does not require direct contact with the circuit?

1. What type of voltage test instrument does not require direct contact with the circuit?

2. Would a quality control testing lab be more likely to use permanent or portable test instruments?

T   F   3. A 1% current unbalance can create an 8% voltage unbalance.

4. Would a maintenance technician who troubleshoots production equipment be more likely to use permanent or portable test instruments?

T   F   5. In-line ammeters are used to measure small amounts of current.

6. Which common multimeter mode records a measurement and displays the difference between the reading and subsequent measurements?

7. Which type of ammeter is used when the circuit to be measured cannot be opened?

8. Which multimeter test mode is commonly used to test components such as switches, fuses, electrical connections, and individual conductors?

T   F   9. The measurement hold mode is a multimeter mode that captures and stores the lowest and highest measurements for later display.

10. Which electrical property equals current multiplied by resistance?

T   F   11. Loads such as motors and solenoids decrease in resistance over time.

12. Which electrical property is measured to indicate total circuit loading?

13. What color is the test lead that typically indicates negative or ground?

T   F   14. Reverse the polarity of the test leads when a negative sign is displayed at the beginning of a multimeter bar graph.

15. Which common multimeter mode would be most useful in measuring the change in current when a load is turned ON?

16. Which common multimeter mode would be most useful in recording high- or low-voltage conditions over a period of time?

17. What type of electronic component is tested by measuring the voltage drop across it in two directions?

T F **18.** Surface leakage current is current that flows from areas on conductors where insulation has been removed to allow for electrical connections.

_____ **19.** Which type of AC voltage measurement is useful because it equates the heat produced by an AC sine wave to DC resistive circuit?

_____ **20.** Which part of an analog multimeter display is determined by the position of the range switch?

21. Why is a small amount (microamperes) of leakage current unavoidable?

22. Why are neon test lights not used to test ground fault circuit interrupters (GFCIs)?

23. Why is it important to test insulation on conductors?

24. Why must megohmmeters, and not regular ohmmeters, be used to test insulation?

25. How is the ohmmeter resistance reading of one component affected by other components connected in parallel?

Application 2-1   *Testing Receptacles*

Test instruments are used to test electrical components and circuits to ensure they are properly installed and working correctly. Even a correctly installed component may not be working properly if there is a component malfunction or the component is not correctly sized for the load or system. After electrical components and circuits are working properly, faults can develop over time. To find the fault, test instruments can be used to troubleshoot the individual components and circuit.

Using test instruments to test a new circuit or troubleshoot a faulty circuit or component requires the knowledge of how to properly set and connect the test instrument into the circuit or to the component and then knowing what the measured values actually mean.

In each of the following applications, a test instrument is used to test a 120 VAC/15 A rated receptacle (outlet). The acceptable voltage range for receptacles is –10% to +5%. See Receptacles.

RECEPTACLES

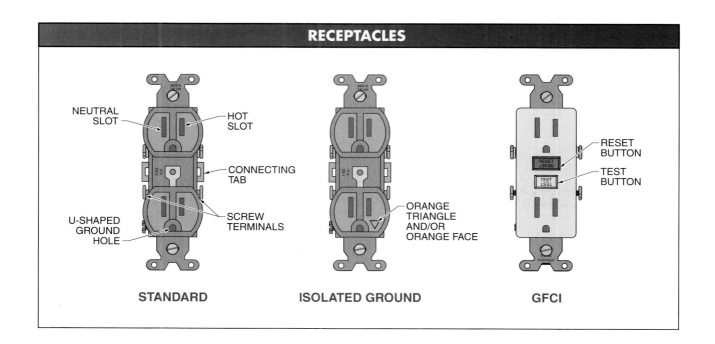

NEUTRAL SLOT

HOT SLOT

CONNECTING TAB

U-SHAPED GROUND HOLE

SCREW TERMINALS

STANDARD

ORANGE TRIANGLE AND/OR ORANGE FACE

ISOLATED GROUND

RESET BUTTON

TEST BUTTON

GFCI

Testing Receptacles with a Test Light

1. Connect the black test lead of Test Light 1 to the neutral side of the receptacle and the red test lead to the hot side of the receptacle.

2. _____ If the receptacle and circuit are working properly, should Test Light 1 be ON or OFF?

3. Connect the black test lead of Test Light 2 to the grounded side of the receptacle and the red test lead to the hot side of the receptacle.

4. _____ If the receptacle and circuit are working properly, should Test Light 2 be ON or OFF?

5. Connect the black test lead of Test Light 3 to the grounded side of the receptacle and the red test lead to the neutral side of the receptacle.

6. _____ If the receptacle and circuit are working properly, should Test Light 3 be ON or OFF?

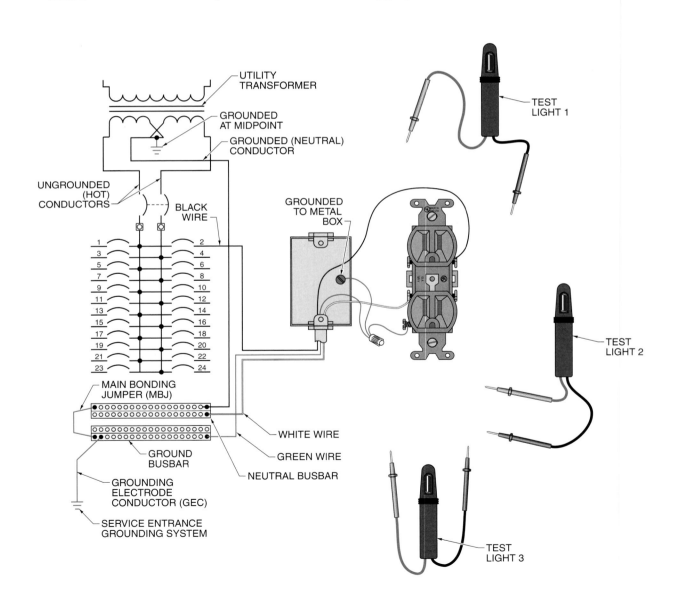

Testing Receptacles with a Voltage Indicator

7. _____ Voltage Indicator 1 is held near one of the slots of Receptacle A. If the receptacle and circuit are working properly, should Voltage Indicator 1 tip be glowing or not glowing?

8. _____ Voltage Indicator 2 is held near one of the slots of Receptacle B. If Voltage Indicator 2 tip is glowing, is the receptacle wired correctly?

9. If Receptacle B is not wired correctly, what is the problem with the circuit?

10. _____ Voltage Indicator 3 is held near one of the slots of Receptacle B. If the receptacle and circuit are working properly, should Voltage Indicator 3 tip be glowing or not glowing?

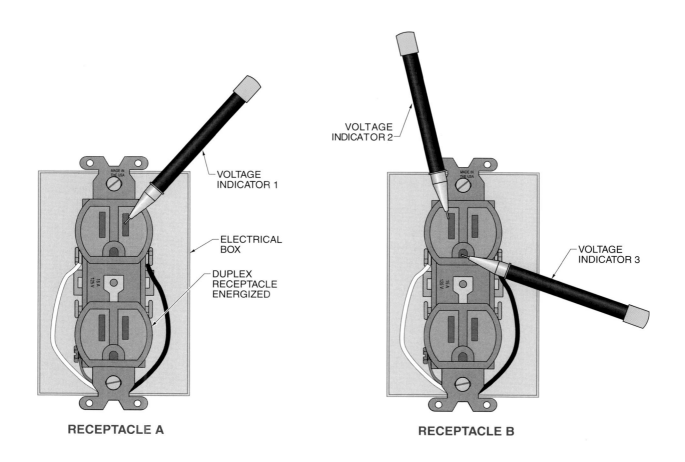

VOLTAGE INDICATOR 1

ELECTRICAL BOX

DUPLEX RECEPTACLE ENERGIZED

VOLTAGE INDICATOR 2

VOLTAGE INDICATOR 3

RECEPTACLE A

RECEPTACLE B

Testing Receptacles with a Voltage Tester

11. Connect the black test lead of Voltage Tester 1 to the neutral side of the GFCI receptacle and the red test lead to the hot side of the receptacle.

12. _____ If the GFCI receptacle and circuit are working properly, what should Voltage Tester 1 be displaying?

13. Connect the black test lead of Voltage Tester 2 to the grounded side of the GFCI receptacle and the red test lead to the hot side of the receptacle.

14. _____ If the GFCI receptacle and circuit are working properly, what should happen when Voltage Tester 2 is connected to the circuit?

15. _____ If the GFCI receptacle and circuit are working properly, what should Voltage Tester 3 be displaying?

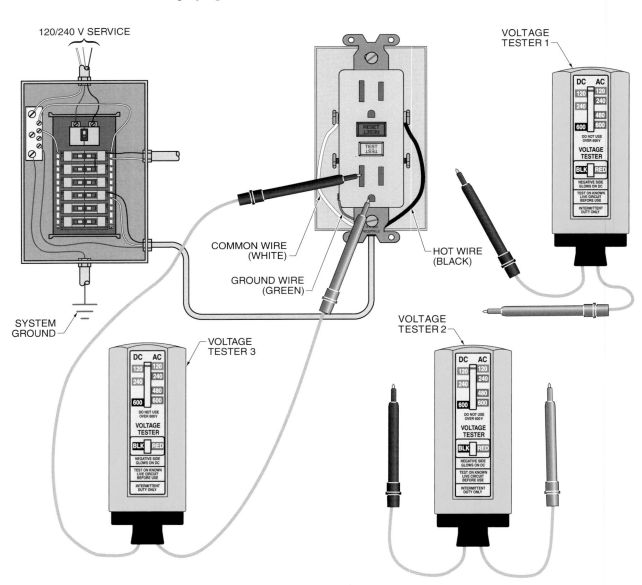

Testing Receptacles with a Basic Multimeter

16. Set (draw in) the multimeter function switch to take voltage measurements at the receptacle.

17. Connect the two test leads to the correct multimeter jacks.

18. Connect the black test lead to the neutral side of the receptacle and the red test lead to the hot side of the receptacle.

19. _____ If the receptacle and circuit are working properly, what is the minimum voltage measurement that should be measured for the circuit to be in the standard acceptable voltage range?

20. _____ If the receptacle and circuit are working properly, what is the maximum voltage measurement that should be measured for the circuit to be in the standard acceptable voltage range?

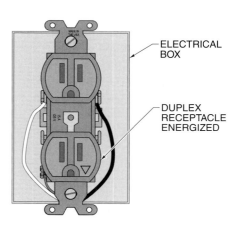

ELECTRICAL BOX

DUPLEX RECEPTACLE ENERGIZED

Testing Receptacles with an Analog Multimeter

21. Set (draw in) the analog multimeter function and range switches to take voltage measurements at the receptacle.

22. Connect the two test leads to the correct jacks on the analog multimeter.

23. Connect the black test lead to the ground side of the receptacle and the red test lead to the hot side of the receptacle.

24. Draw the position that the pointer of the analog multimeter should be at if the circuit is wired correctly.

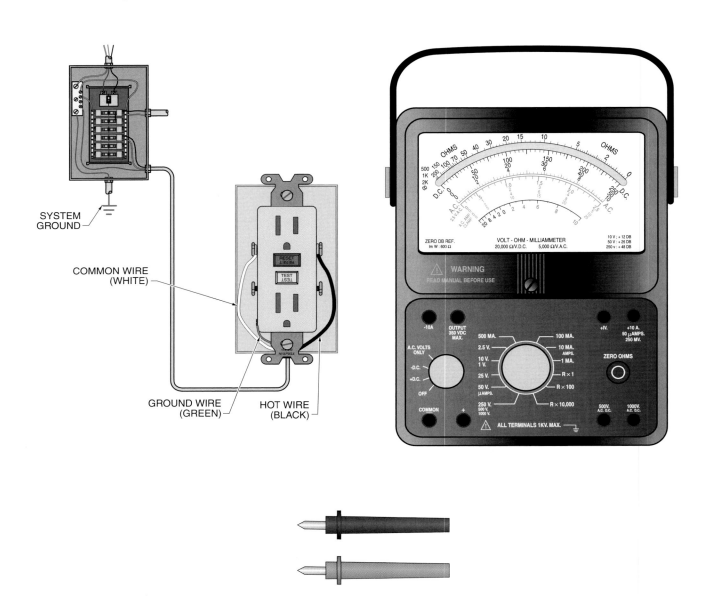

Testing Receptacles with an Advanced Multimeter

25. Set (draw in) the multimeter function switch to take voltage measurements at the receptacle.

26. Connect the two test leads to the correct jacks on the multimeter.

27. Connect the black test lead to the neutral side of the receptacle and the red test lead to the hot side of the receptacle.

28. If measurements are to be taken over time to see if the voltage in the circuit is fluctuating, circle the multimeter function button that would be used.

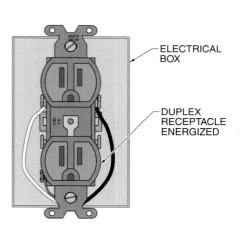

ELECTRICAL BOX

DUPLEX RECEPTACLE ENERGIZED

Test Instrument Abbreviations and Symbols

All test instruments use abbreviations and symbols to simplify electrical terms. Understanding the abbreviations and symbols used with test instruments is required in order to properly set and use test instruments. Identify each abbreviation and symbol used on each test instrument.

Ground Resistance Meter

1. _____   Symbol A = ___.

2. _____   Symbol B = ___.

3. _____   Symbol C = ___.

4. _____   Symbol D = ___.

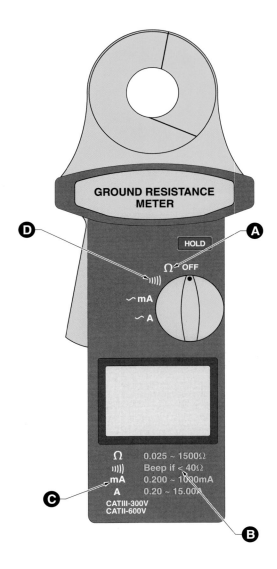

Analog Multimeters

5. _____ Symbol A = ___.

6. _____ Symbol B = ___.

7. _____ Symbol C = ___.

8. _____ Symbol D = ___.

9. _____ Symbol E = ___.

10. _____ Symbol F = ___.

Digital Multimeters

11. _____ Symbol A = ___.

12. _____ Symbol B = ___.

13. _____ Symbol C = ___.

14. _____ Symbol D = ___.

15. _____ Symbol E = ___.

16. _____ Symbol F = ___.

17. _____ Symbol G = ___.

18. _____ Symbol H = ___.

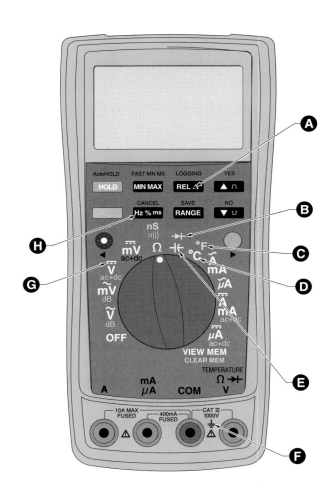

Analog multimeters have several calibrated scales that correspond to different range switch settings. When setting up and reading a measurement on an analog multimeter, the position of the range switch must be set at the correct quantity, and the correct scale must be read. The range switch is set to a voltage setting (2.5 V, 10 V, 25 V, etc.) position when measuring voltage (V). The range switch is set to a resistance setting (R × 1, R × 100, R × 10,000) position when measuring resistance (R). The range switch is set to a current setting (1 mA, 10 mA, 100 mA, etc.) position when measuring current (mA or A). See Range Switch Settings and Analog Scales.

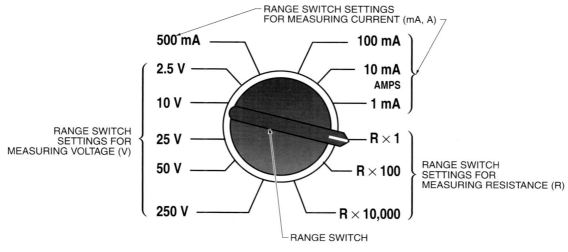

RANGE SWITCH SETTINGS

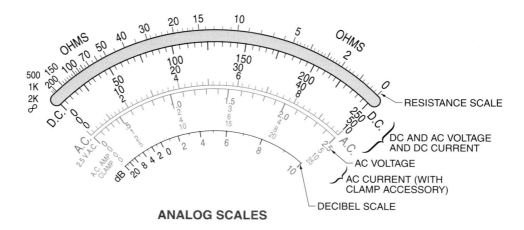

ANALOG SCALES

Each scale has multiple uses depending on the range switch setting. For example, the 0 to 50 scale can also be used to measure 0 to 500 by multiplying each number by 10. This scale can also be used to measure DC or AC, and voltage or current.

To prevent damage to the meter's moving pointer, the highest voltage and current range should be used first when taking a voltage or current measurement. The range switch can be adjusted if the reading is below a lower range setting. This will allow a more accurate reading of the analog scale. A high resistance measurement taken on a low resistance meter range switch setting, however, does not damage the meter. In this situation, the pointer will just move to the infinity (∞) resistance position.

The function switch is set on AC when alternating voltage (VAC) is measured, and is set on DC when direct voltage or current (DC) is measured. If an analog multimeter does not have a separate setting for resistance, either DC setting is typically used. See Function Switch.

The normal setting for measuring DC voltage or current is +DC. This setting makes the red test lead positive. The alternative setting for measuring DC voltage or current is –DC. This setting makes the red test lead negative. The –DC setting is not common, but is used on some DC circuits that use a battery's positive terminal as the common ground.

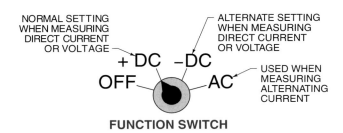

FUNCTION SWITCH

List the correct reading for the range settings.

1. _____   Reading = ___ VDC

2. _____   Reading = ___ mA (DC)

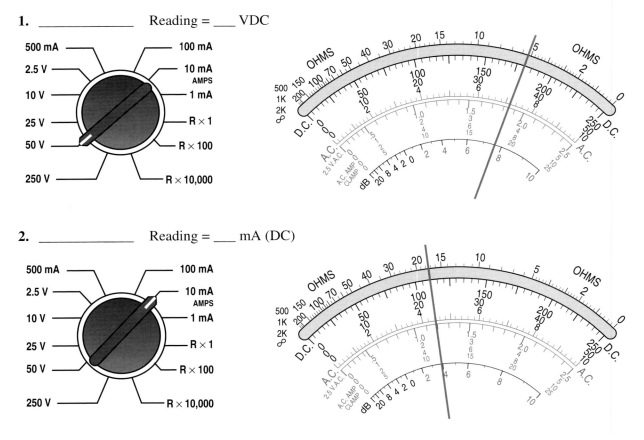

3. _____ Reading = ___ kΩ

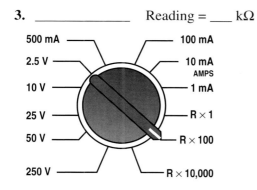
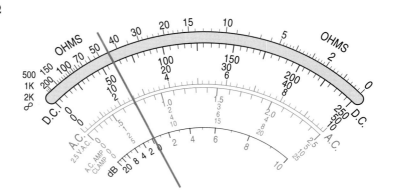

4. _____ Reading = ___ VAC

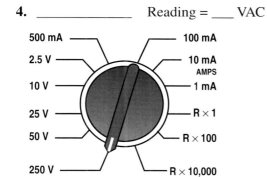
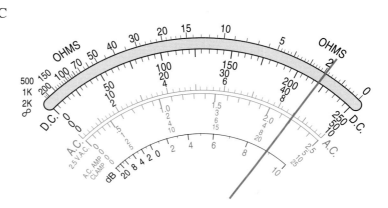

5. _____ Reading = ___ mA (DC)

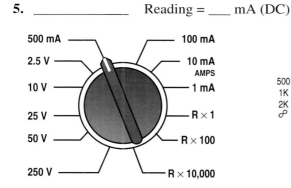
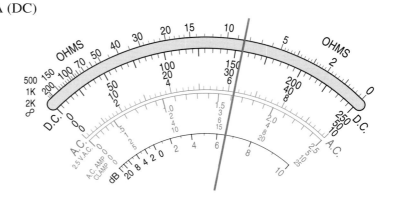

6. _____ Reading = ___ Ω

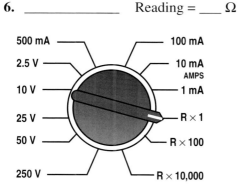
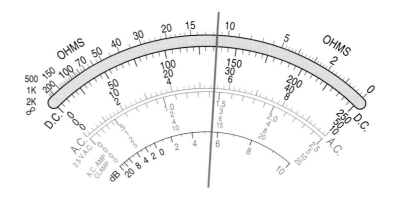

Troubleshooting Switches

Test instruments are used to take measurements to verify that a circuit or component is working properly. Before a test instrument is connected into a circuit, the test instrument must be set properly. When connecting a test instrument into a circuit, approximate test instrument readings should be anticipated if the test instrument readings are going to be used to help determine circuit problems. Test instrument readings that are not understood cannot aid in understanding the circuit or finding a problem.

1. Set (draw in) the function switch on Multimeter 1 to measure voltage at the switch.

2. Set (draw in) the function switch on Multimeter 2 to measure voltage at the lamp.

3. _____ If the circuit is working properly, what should Multimeter 1 read when the two-way switch is in the open (OFF) position?

4. _____ If the circuit is working properly, what should Multimeter 2 read when the two-way switch is in the open (OFF) position?

5. _____ If the circuit is working properly, what should Multimeter 1 read when the two-way switch is in the closed (ON) position?

6. _____ If the circuit is working properly, what should Multimeter 2 read when the two-way switch is in the closed (ON) position?

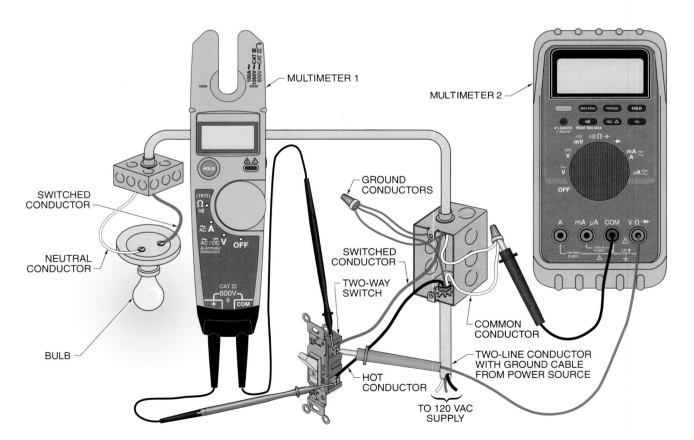

Application 2-5

Understanding Current Measurements

Current measurements are taken when testing or troubleshooting a circuit. Current measurements are used to determine how much load is on a circuit. Current measurements should be taken over time or at different operating times because current can vary as a circuit's load condition changes (loads turned ON and OFF, and/or motor loads change).

The incoming power supply and motor are configured for 220 VAC.

1. Set (draw in) the function switch on Multimeters 1, 2, and 3 to measure the current of the motor.

2. _____ What is the expected current measurement if the motor is fully loaded to the nameplate horsepower rating?

3. _____ What is the expected current measurement if the motor is fully loaded to the nameplate horsepower rating and operating at its service factor rating?

4. _____ What is the expected current measurement if the motor is operating at 80% of its full load rating?

5. _____ If the incoming power supply and motor are wired for 440 VAC, what is the expected current measurement if the motor is fully loaded to the nameplate horsepower rating?

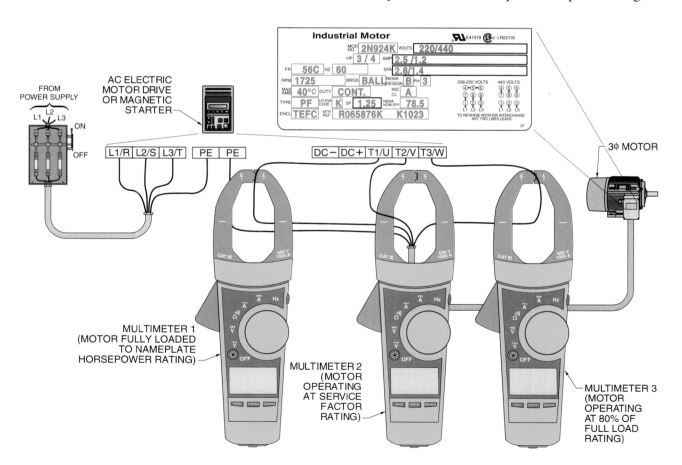

All motors have motor windings (conductors) that develop magnetic fields that produce rotation when connected to power. All motor winding conductors have resistance. The more horsepower a motor produces, the larger the size of winding conductor required. The larger the size of the conductor, the smaller the AWG number. The larger the conductor, the less the resistance. Taking resistance measurements and applying the laws of resistance in series and parallel circuits can help verify connections to determine if a motor is wired correctly before any power is applied to the motor.

Resistance in Series Circuits

The total resistance in a circuit containing series-connected components (motor windings) equals the sum of the resistances of all components. The resistance in the circuit increases if components (motor windings) are added in series and decreases if components are removed. To calculate total resistance in a series circuit, apply the following formula:

$$R_T = R_1 + R_2 + R_3 + \ldots$$

where

R_T = total circuit resistance (in Ω)

R_1 = resistance 1 (in Ω)

R_2 = resistance 2 (in Ω)

R_3 = resistance 3 (in Ω)

For example, what is the total resistance in a circuit that has 2 Ω, 4 Ω, and 6 Ω resistors connected in series?

$$R_T = R_1 + R_2 + R_3$$
$$R_T = 2 + 4 + 6$$
$$R_T = \textbf{12} \; \boldsymbol{\Omega}$$

Resistance in Parallel Circuits

The total resistance in a circuit containing parallel-connected components (motor windings) is less than the smallest resistance value. The total resistance decreases if loads are added in parallel and increases if loads are removed. To calculate total resistance in a parallel circuit containing two resistors, apply the following formula:

$$R_T = \frac{R_1 \times R_2}{R_1 + R_2}$$

where

R_T = total circuit resistance (in Ω)

R_1 = resistance 1 (in Ω)

R_2 = resistance 2 (in Ω)

For example, what is the total resistance in a circuit containing resistors of 16 Ω and 24 Ω connected in parallel?

$$R_T = \frac{R_1 \times R_2}{R_1 + R_2}$$

$$R_T = \frac{16 \times 24}{16 + 24}$$

$$R_T = \frac{384}{40}$$

$$R_T = \mathbf{9.6\ \Omega}$$

The total resistance of more than two resistors with different values, connected in parallel is calculated by applying the following formula:

$$R_T = \frac{1}{\dfrac{1}{R_1} + \dfrac{1}{R_2} + \dfrac{1}{R_3} + \cdots}$$

where

R_T = total circuit resistance (in Ω)

R_1 = resistance 1 (in Ω)

R_2 = resistance 2 (in Ω)

R_3 = resistance 3 (in Ω)

For example, what is the total resistance in a circuit containing resistors of 16 Ω, 24 Ω, and 48 Ω connected in parallel?

$$R_T = \frac{1}{\dfrac{1}{R_1} + \dfrac{1}{R_2} + \dfrac{1}{R_3}}$$

$$R_T = \frac{1}{\dfrac{1}{16} + \dfrac{1}{24} + \dfrac{1}{48}}$$

$$R_T = \frac{1}{0.0625 + 0.0417 + 0.0208}$$

$$R_T = \frac{1}{0.125}$$

$$R_T = \mathbf{8\ \Omega}$$

Testing Single-Voltage, Wye-Connected, Three-Phase Motor Wiring

The motor winding resistance for Motor Coil T1 to internal connection point is 4 Ω before the motor leads are connected (as given by the motor manufacturer data). After the motor leads are connected to the motor side of the motor starter, resistance measurements are taken before power is applied to verify that the motor is connected correctly.

1. _____ What is the resistance between T1 and T2?

2. _____ What is the resistance between T2 and T3?

3. _____ What is the resistance between T3 and T1?

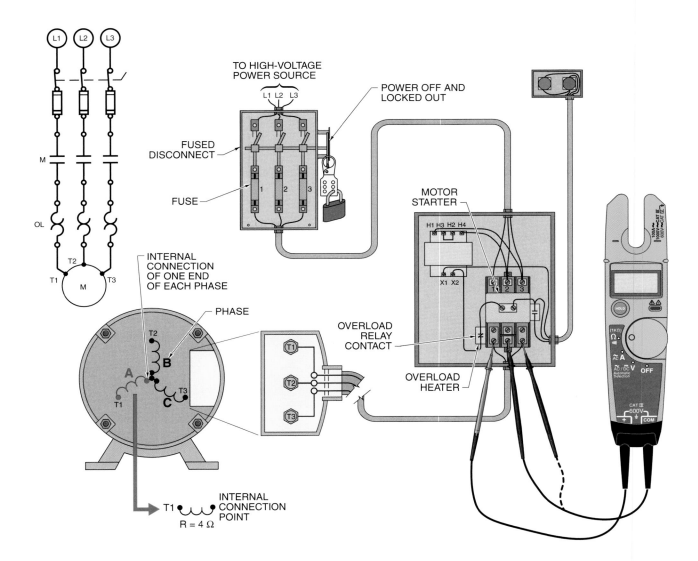

Testing Single-Voltage, Delta-Connected, Three-Phase Motor Wiring

The motor winding resistance for Motor Coil A is 6 Ω before the motor internal leads are connected (as given by the motor manufacturer data). After the motor leads are connected to the motor side of the motor starter, resistance measurements are taken before power is applied to verify that the motor is connected correctly.

4. _____ What is the resistance between T1 and T2?

5. _____ What is the resistance between T2 and T3?

6. _____ What is the resistance between T3 and T1?

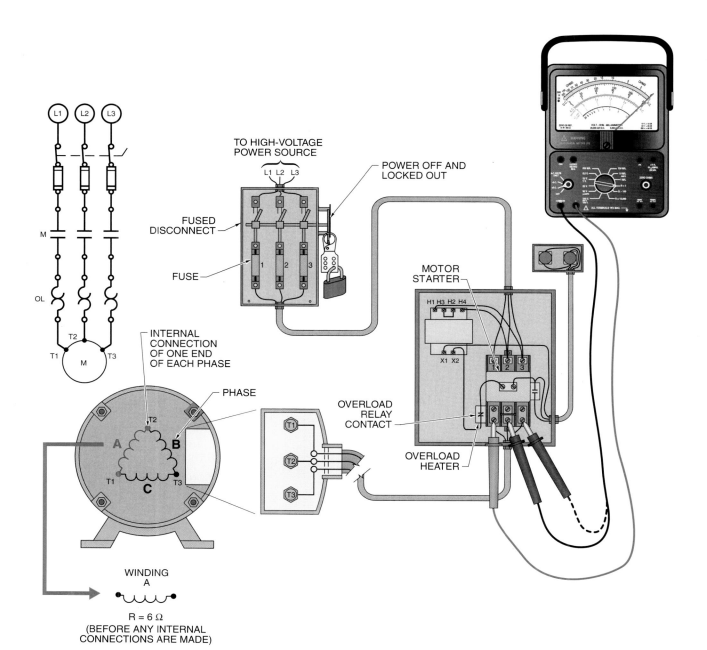

Testing Dual-Voltage, Wye-Connected, Three-Phase Motor Wiring

The motor winding resistance for Motor Coil T1 to T4 is 12 Ω before the motor leads are connected. After the motor leads are connected to the motor side of the motor starter, resistance measurements are taken before power is applied to verify that the motor is connected correctly.

7. _____ What is the resistance between T1 and T2?

8. _____ What is the resistance between T2 and T3?

9. _____ What is the resistance between T3 and T1?

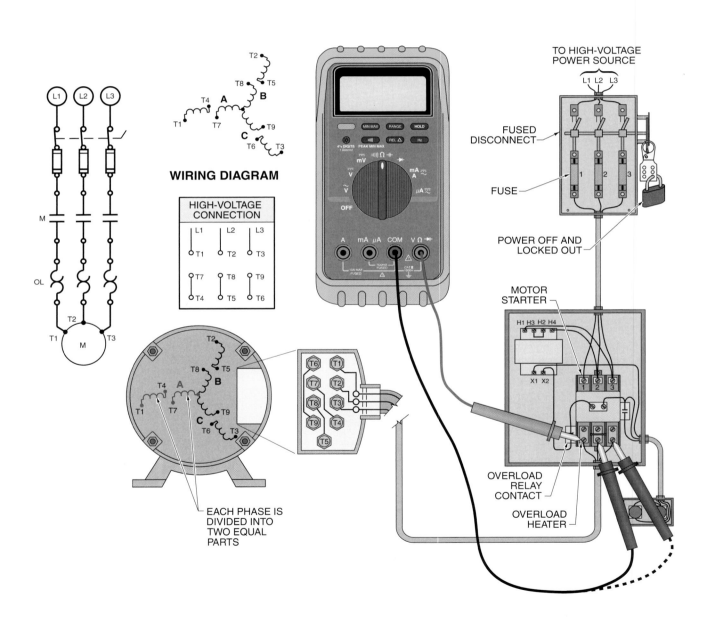

WIRING DIAGRAM

HIGH-VOLTAGE CONNECTION		
L1	L2	L3
T1	T2	T3
T7	T8	T9
T4	T5	T6

EACH PHASE IS DIVIDED INTO TWO EQUAL PARTS

TO HIGH-VOLTAGE POWER SOURCE

L1 L2 L3

FUSED DISCONNECT

FUSE

POWER OFF AND LOCKED OUT

MOTOR STARTER

H1 H3 H2 H4

X1 X2

OVERLOAD RELAY CONTACT

OVERLOAD HEATER

Testing Dual-Voltage, Delta-Connected, Three-Phase Motor Wiring

The motor winding resistance for Motor Coil T1 to T4 is 10 Ω before the motor leads are connected. After the motor leads are connected to the motor side of the motor starter, resistance measurements are taken before power is applied to verify that the motor is connected correctly.

10. _____ What is the resistance between T1 and T2?

11. _____ What is the resistance between T2 and T3?

12. _____ What is the resistance between T3 and T1?

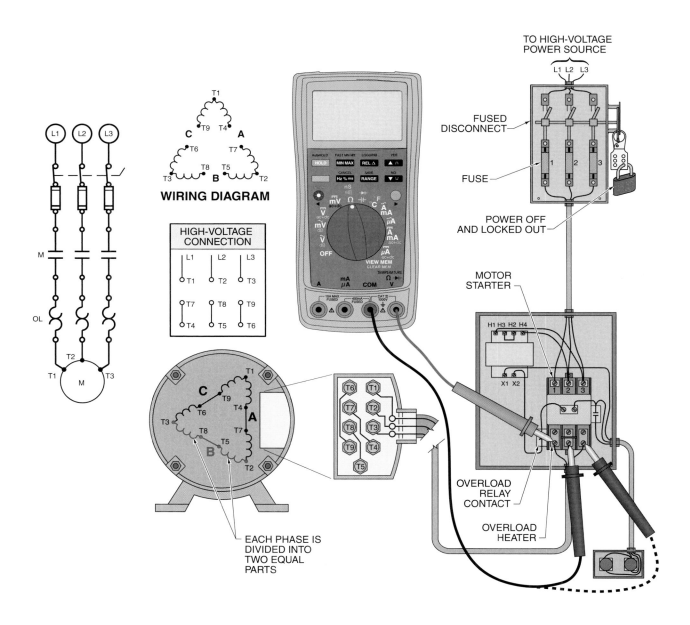

Application 2-7 — *Selecting Appropriate Test Instruments*

Some test instruments are designed for one specific purpose while others are designed for a range of tasks. Every test instrument has advantages and disadvantages that determine its usefulness in a given situation. Selecting the appropriate test instrument for a particular measurement or troubleshooting application requires an understanding of which electrical quantity needs to be measured (volts, amps, ohms, etc.), how the measured quantity is to be displayed (maximum value, numerical value, waveform, etc.), and the limits of the test instrument (CAT rating, current limit, etc.).

In general, the least complicated piece of test equipment that can perform the required task is recommended. For example, a test light, receptacle tester, voltage tester, circuit analyzer, and multimeter can all be used to check a receptacle (outlet). If only an indication of whether the receptacle is powered (live) is needed, the test light will work. If the receptacle must be tested for power and correct wiring, the receptacle tester will work. However, if more advanced measurements or tests (voltage levels, voltage recorded over time, peak voltage, etc.) are required at the receptacle, then a voltage tester, circuit analyzer, or multimeter must be used.

All limits on voltage and current measurements (whether the current jacks are fused or unfused, the condition of the test leads, etc.) must be considered as well as factors such as the CAT ratings of the application and the test instrument.

Select a test instrument from the following that best meets the measurement requirement for each situation. Also, give the reason for the selection. The simplest test instrument that can perform the required task should be selected, even if more than one test instrument could be used.

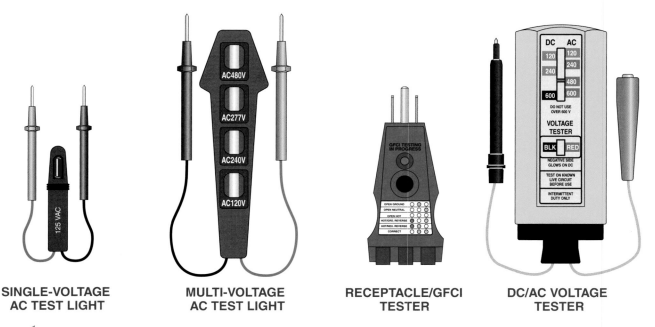

SINGLE-VOLTAGE AC TEST LIGHT **MULTI-VOLTAGE AC TEST LIGHT** **RECEPTACLE/GFCI TESTER** **DC/AC VOLTAGE TESTER**

SIMPLEST

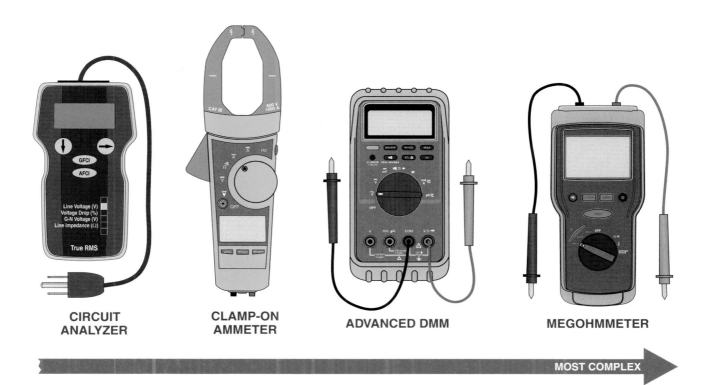

| CIRCUIT ANALYZER | CLAMP-ON AMMETER | ADVANCED DMM | MEGOHMMETER |

MOST COMPLEX

A service call requests that several receptacle circuits in an office be checked at the branch circuit power panel at a high usage time to see if any of the circuits are near their current limit (rating of the circuit breaker).

1. _____ Which is the simplest test instrument that meets the service call requirements?

2. Why is this test instrument the best choice?

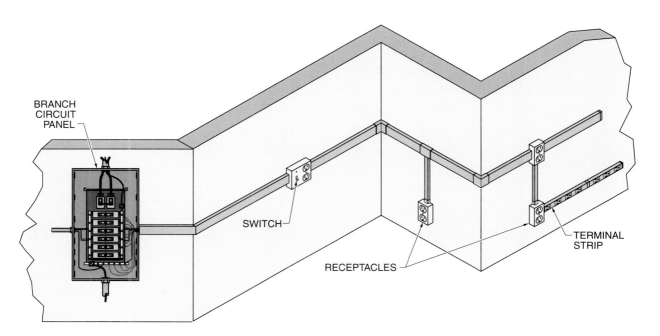

A service call requests that all the receptacles in the office (including the terminal strip) be checked to make sure they are powered and grounded, and that any GFCI receptacles be tested before office computers are plugged into them.

3. _____ Which is the simplest test instrument that meets the service call requirements?

4. Why is this test instrument the best choice?

When the office equipment is connected, the receptacles should be retested to compare the rms voltage to the peak voltage. (The peak voltage should be 1.414 times the rms voltage.) The line impedance should also be tested to make sure the branch circuit conductors are not undersized and the circuit runs are not too long.

5. _____ Which is the simplest test instrument that meets the service call requirements?

6. Why is this test instrument the best choice?

A service call states that some computers are automatically rebooting during high-usage times when printers and fax machines on the same circuit are operating. The voltage should be measured and recorded for one complete working day (10 hr) to determine the lowest voltage during that time.

7. _____ Which is the simplest test instrument that meets the service call requirements?

8. Why is this test instrument the best choice?

A service call requests that a switch be tested to determine if it controls the receptacle next to the switch. It may control both sides, only one side, or not control the receptacle at all.

9. _____ Which is the simplest test instrument that meets the service call requirements?

10. Why is this test instrument the best choice?

_____ **1.** What does UTP stand for?

_____ **2.** What is the function of an outer braided conductor in a cable?

_____ **3.** What does the CAT rating of a conductor indicate?

T   F **4.** Fiber optic cable uses light signals to transmit communications data.

_____ **5.** Which VDV test measures the power loss of a signal over a cable pair?

_____ **6.** Which type of communication signal does not require hard wiring between locations?

T   F **7.** Signal losses from splices and connectors often exceed losses from cable length.

_____ **8.** What are the two conductors in a twisted pair called?

_____ **9.** Does a lower CAT rating indicate a higher or lower data transfer capacity?

_____ **10.** Which VDV test measures the time required for a signal to travel the length of the cable pair?

T   F **11.** The color code for the first pair in a communication cable typically includes red.

_____ **12.** Should communications cables be run in parallel with power conductors?

_____ **13.** Which VDV test uses voice signals to verify communication circuits?

T   F **14.** Fiber optic signals are susceptible to EMI and RFI.

T   F **15.** Coaxial cable typically carries video signals.

_____ **16.** Which VDV cable problem results in signals in one circuit inducing unwanted signals in another circuit?

_____ **17.** Which test instrument can be used to visually examine contaminants or faults in fiber optic cables?

_____ **18.** Which VDV test determines if wires in communications circuits are crossed, reversed, open, shorted, or split?

T   F **19.** 10 dB of power loss is equivalent to 10% loss.

_____ **20.** Which VDV test measures the total opposition to alternating current on a cable?

21. What are some possible sources of electromagnetic or radio frequency interference?

22. What is the advantage to twisting cable pairs?

23. Explain how a tone generator and amplifier probe work together to test communication wiring.

24. What advantages do fiber optic networks have over copper networks?

25. How does a time domain reflectometer (TDR) measure the distance to a cable fault or the end of a cable?

Name_____ Date_____

Cable Insulation Testing

Conductors are coated with insulation to prevent contact with other conductors, metal parts, and people. Unwanted contact results in electrical shorts that can cause communications failure, equipment damage, and hazardous situations. Insulation prevents electrical shorts because it is a high-resistance material that resists electron flow. Insulation resistance is reduced by physical damage from cuts, crimps, or stress, and deterioration from age, heat, UV light, or moisture. Since cables are usually inside hidden cable runs, damaged insulation is not always easily found. The resistance of communication cable insulation is measured by testing the conductors in a run alongside each other and ground using a megohmmeter. See Insulation Resistance Testing.

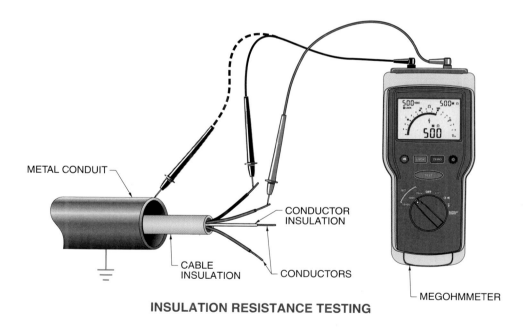

METAL CONDUIT

CONDUCTOR INSULATION

CABLE INSULATION

CONDUCTORS

MEGOHMMETER

INSULATION RESISTANCE TESTING

The resistance of cable insulation should be tested when cables are installed, as part of a predictive maintenance schedule (every couple of years), and when there is a problem in the system that has not been solved. Older systems should be checked more frequently because aging accelerates the deterioration of insulation.

All insulation resistance tests are made with power OFF to cables and cables disconnected at both ends from any devices. For low voltage cables (rated for 1000 V or less), the megohmmeter applies 1000 VDC for 1 min to measure resistance. The resistance measured from conductor to conductor and conductor to ground must be a minimum of several megohms and will probably measure hundreds of megohms on good insulation.

The true measure of the condition of insulation is obtained by taking measurements over time, such as annually. Small changes in resistance value over time are acceptable. For example, a change from 300 MΩ to 275 MΩ is only about an 8.3% change and does not indicate a problem. However, a change from 300 MΩ to 200 MΩ is a 33% change and indicates a problem. See Insulation Resistance Changes.

INSULATION RESISTANCE CHANGES*												
Baseline Measurement	**20**	**40**	**60**	**80**	**100**	**250**	**500**	**750**	**1000**	**1500**	**2000**	**Difference Interpretation**
10	18	36	54	72	90	225	450	675	900	1350	1800	Acceptable operating range for most equipment
20	16	32	48	64	80	200	400	600	800	1200	1600	
30	14	28	42	56	70	175	350	525	700	1050	1400	Requires additional testing and inspection to verify proper operation; check for environmental contamination
40	12	24	36	48	60	150	300	450	600	900	1200	
50	10	20	30	40	50	125	250	375	500	750	1000	
60	8	16	24	32	40	100	200	300	400	600	800	Indicates a potential problem; perform tests on system until problem is located and corrected
70	6	12	18	24	30	75	150	225	300	450	600	
80	4	8	12	16	20	50	100	150	200	300	400	
90	2	4	6	8	10	25	50	75	100	150	200	
100	0	0	0	0	0	0	0	0	0	0	0	

% Difference from Baseline

* in MΩ

Note: Baseline measurements are used as comparison readings for field measurements. A baseline measurement is typically the manufacturer's factory specifications or the last recorded reading from a preventive maintenance procedure.

1. _____ What is the percentage change between the first and second reading?

2. _____ Is this percentage change acceptable?

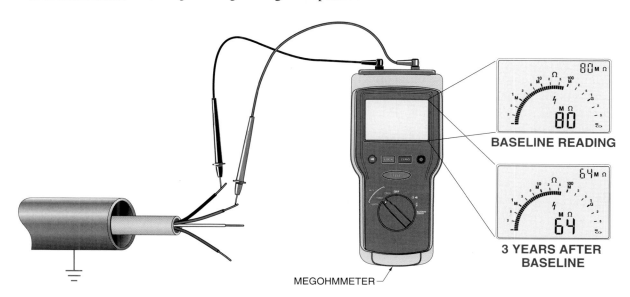

3. _____ What is the percentage change between the first and second reading?

4. _____ Is this percentage change acceptable?

5. _____ What is the percentage change between the second and third reading?

6. _____ Is this percentage change acceptable?

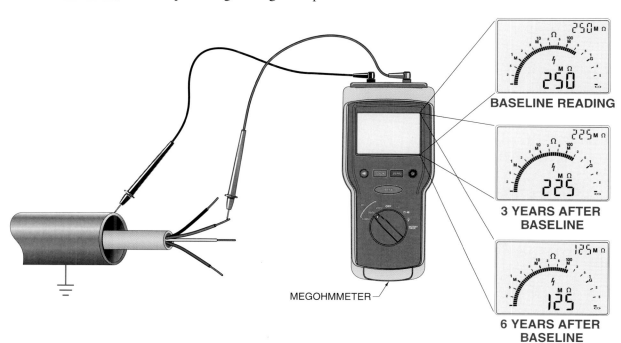

When experiencing a communications problem, one of the first tests to perform is a wire map test, which checks the order and arrangement of wires at two connection points. Wire map testers determine if a cable contains a fault or wiring error, or is wired correctly. See Correct Wiring. A fault is a short or open and is usually caused by either physical damage to the wire or incomplete terminations at the connectors. Wiring errors include three types of incorrect wire arrangements. A reversed pair has the two wires of a pair reversed at one end. A set of crossed pairs has the two pairs reversed at one end. A set of split pairs forms two pairs at each end by mixing one wire from each of two twisted pairs. See Wiring Errors.

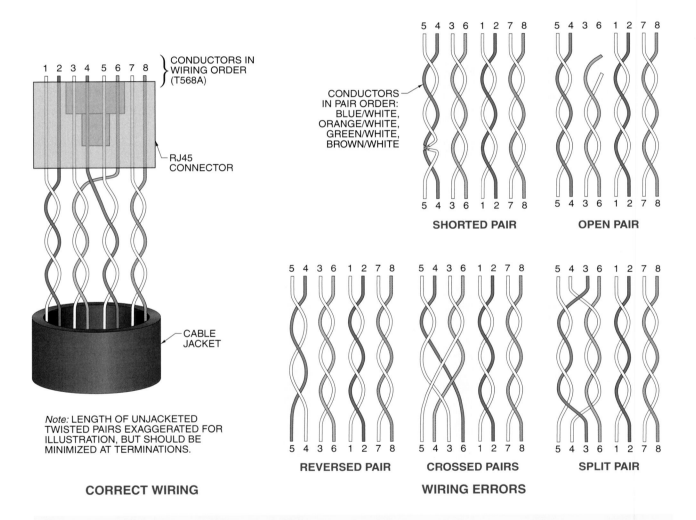

CORRECT WIRING

WIRING ERRORS

Wire map tests can be performed by individual testers or by more advanced cable analyzers that include wire mapping as one of the several tests they perform. The tester may require a separate unit to be placed at the opposite end of the cable from the tester to help map the wires at both points. When interpreting the results on the meter display, be sure to note the wire numbers. Some meters may display the pairs in different orders.

1. _____ What type of error does Wire Map Test 1 show?

2. _____ What type of error does Wire Map Test 2 show?

3. How could the error in Wire Map Test 2 be corrected?

4. _____ What type of error does Wire Map Test 3 show?

5. How could the error in Wire Map Test 3 be corrected?

6. _____ What type of error does Wire Map Test 4 show?

7. _____ What type of error does Wire Map Test 5 show?

8. How could the error in Wire Map Test 5 be corrected?

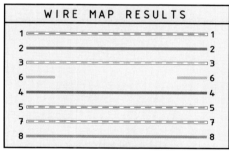

WIRE MAP TEST 1

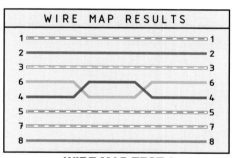

WIRE MAP TEST 2

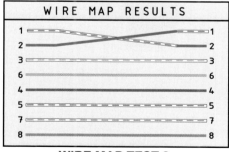

WIRE MAP TEST 3

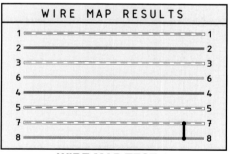

WIRE MAP TEST 4

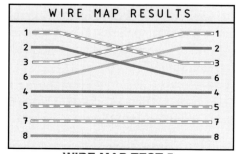

WIRE MAP TEST 5

A time domain reflectometer (TDR) test measures conductor length and locates faults (shorts or opens) and poor connections on conductors. The meter sends out a pulse and measures the time it takes for a reflection of the pulse to return. Reflections are caused by the end of the conductor, faults, and poor connections. By comparing the reflection pulse's travel time, shape, and polarity to the original transmitted pulse, the meter can determine the distance to and cause of the reflection.

A properly terminated conductor with no faults has uniform impedance and absorbs the pulse without reflecting it. A fault in a conductor produces an abrupt change in the conductor's impedance that causes reflections of the pulse to return to the TDR. The amplitude of the reflection is proportional to the change of impedance. An open in the conductor represents an abrupt increase in the conductor's impedance, which sends a reflection of the pulse back to the meter with the same polarity. An unterminated conductor will also reflect the pulse from the end of the conductor, which is like an open, even if the conductor has no faults. See Open Cable. A short in the conductor represents an abrupt decrease in the conductor's impedance, which sends a reflection of the pulse back to the meter with the opposite polarity. See Shorted Cable.

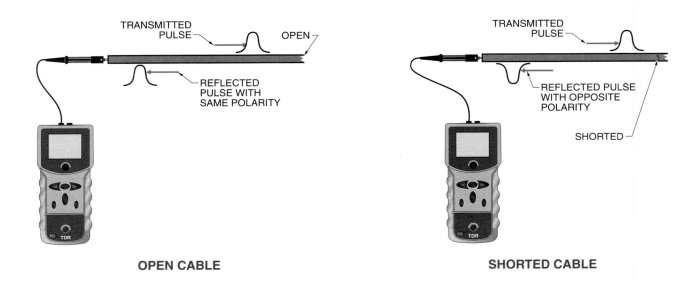

OPEN CABLE   **SHORTED CABLE**

Most TDRs interpret the signals internally and only display the results, which usually include the type of fault (if any) and the distance estimate to the fault. However, some TDRs will display the actual waveform of the pulses. The initial pulse appears as a positive waveform on the display of impedance change versus time. Any reflections appear as subsequent waves of variable amplitudes and either polarity. Partial and complete opens appear as reflections of the same polarity. See TDR Waveform of Opens. Partial and complete shorts appear as reflections of the opposite polarity See TDR Waveform of Shorts.

There are many other types of faults or wiring components that can be detected by analyzing a TDR waveform, including water infiltration, splices, taps, and radio frequency (RF) interference. Bad connections are relatively common faults and appear as a high impedance reflection followed closely by a low impedance reflection, forming an "S" shape. See TDR Waveform of Bad Connection.

The physical location of a fault can be calculated from its reflection's travel time (using the time scale). Many TDRs will calculate distances internally and show a result for the location of the display curser, or just substitute the time scale for a distance scale (typically in meters). The location can also be approximated when the waveform shows the end of the cable. For example, if a waveform shows a partial short about halfway between the initial pulse and a complete open (the end of an unterminated cable), then the partial short is located about halfway down the conductor.

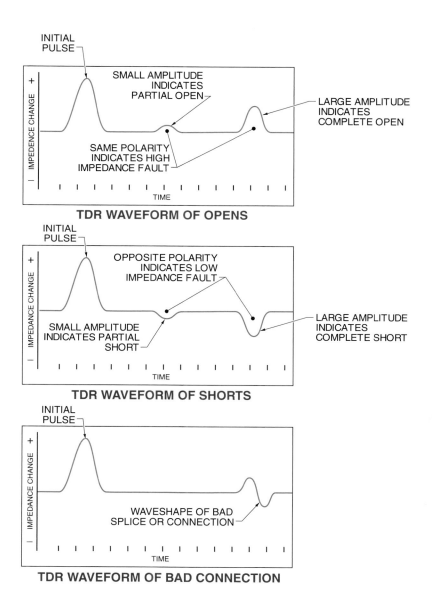

TDR WAVEFORM OF OPENS

TDR WAVEFORM OF SHORTS

TDR WAVEFORM OF BAD CONNECTION

A waveform TDR tests four conductors in a communications cable installation and displays the results as waveforms.

1. _____   What type of fault (if any) is present on Conductor 1?

2. _____   Is Conductor 1 properly terminated at the far end?

3. _____   What type of fault (if any) is present on Conductor 2?

4. _____   Is Conductor 2 properly terminated at the far end?

5. _____   What type of fault (if any) is present on Conductor 3?

6. _____   Is Conductor 3 properly terminated at the far end?

7. _____   What type of fault (if any) is present on Conductor 4?

8. _____   Is Conductor 4 properly terminated at the far end?

9. _____   Approximately how far down Conductor 4 is the fault?

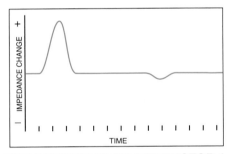

TDR WAVEFORM OF CONDUCTOR 1

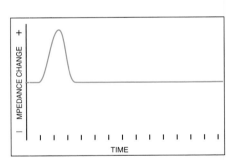

TDR WAVEFORM OF CONDUCTOR 2

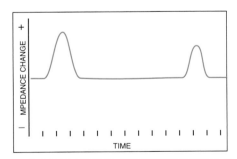

TDR WAVEFORM OF CONDUCTOR 3

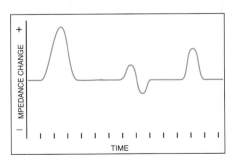

TDR WAVEFORM OF CONDUCTOR 4

Application 3-4

Attenuation

A transmitted signal becomes weaker as it travels down a conductor. If the signal becomes too weak for the system hardware to interpret the signal, the transmitted data is lost. In copper cabling, signal loss occurs because of cable resistance, impedance mismatch, insulation breakdown, crosstalk, and other problems. In fiber optic cable, signal loss is caused by problems including dirty connections, scratches on fiber ends, or misaligned connections. An attenuation test is a test that measures the attenuation (power loss) of cable. Two meters are used to measure a conductor's attenuation. The remote unit is connected to one end of the conductor to transmit a fixed signal and the meter is connected to the other end of the conductor to measure the strength of the transmitted signal.

Attenuation is measured in decibels (dB). For fiber optic cable, the decibel is a logarithmic ratio of output power to input power. For every 3 dB of loss, the output power is approximately one half the input power, and for every 10 dB of loss, the output power is approximately one tenth of the input power.

Decibels are useful because they can be easily added together, as opposed to ratios, which must be multiplied. For example, one length of fiber optic cable attenuates a signal to one half of the input power (–3 dB). Another length attenuates a signal to one tenth of the input power (–10 dB). If a signal travels the length of both cables, only 5% of the input power is left at the end ($\frac{1}{2} \times \frac{1}{10} = \frac{1}{20} = 5\%$). If the two cables are connected together, the total loss from the cables is 95%, which instead can be easily expressed as –13 dB by adding the decibel values together (–3 dB + [–10 dB] = –13 dB). See Compounding Attenuation.

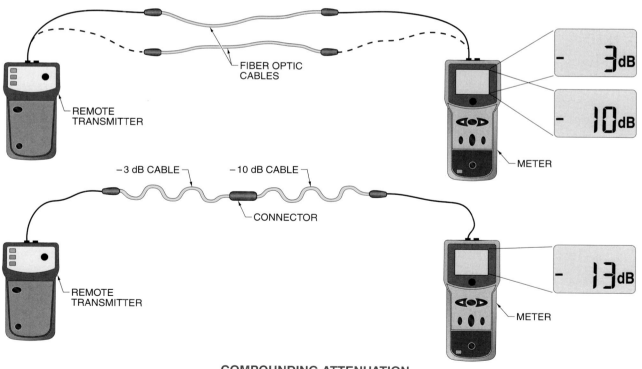

COMPOUNDING ATTENUATION

Incoming data lines can be easily connected to various wall jacks throughout a building by making connections on a patch panel with very short cables. In this scenario, assume the patch cables contribute no additional signal losses.

1. _____ What percentage of power is lost from Wall Jack 1 to the patch panel?

2. _____ What percentage of power is lost from Wall Jack 2 to the patch panel?

3. _____ What percentage of power is lost from Wall Jack 3 to the patch panel?

4. _____ If Wall Jack 1 is patched to the building network, what is the total loss (in dB)?

5. _____ If Wall Jack 1 is patched to the building network, what is the approximate total loss (in %)?

6. _____ If Wall Jack 2 is patched to the building network, what is the total loss (in dB)?

7. _____ If Wall Jack 2 is patched to the building network, what is the approximate total loss (in %)?

8. _____ If Wall Jack 3 is patched to the building network, what is the total loss (in dB)?

9. _____ If Wall Jack 3 is patched to the building network, what is the approximate total loss (in %)?

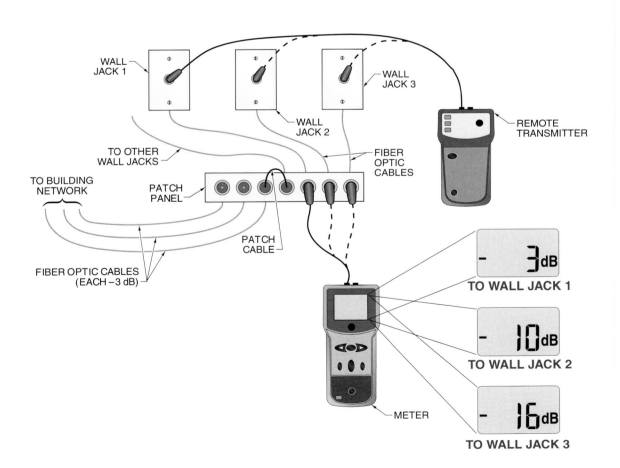

Application 3-5   *Fiber Optic Microscope Views*

Fiber optic microscopes are used to view the endfaces (terminations) of fiber optic cables. To make a proper fiber optic connection, the fiber must be cut cleanly and the endface polished smooth. A good termination will appear as a clear image in a fiber optic microscope, with a light gray core surrounded by darker cladding. See Fiber Optic Microscope.

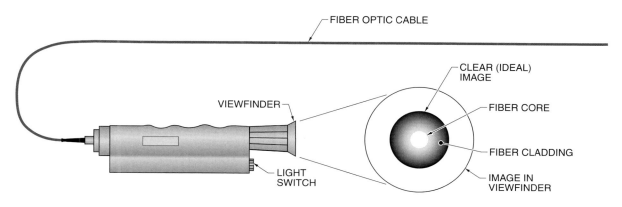

FIBER OPTIC MICROSCOPE

Images with dirt, cracks, scratches, or fuzziness (roughness) indicate problems with the termination. Some problems, such as light scratches or roughness on the end surface of the fiber, can be resolved by cleaning or re-polishing. Others, such as cracks and shattered ends, require the fiber to be cut and terminated again.

Fiber optic microscopes can only be used to view the end of a fiber optic cable. Problems such as chipped, broken, or cracked fibers can also occur anywhere along the fiber cable run, but these require an optical time domain reflectometer (OTDR) to diagnose.

From the following list, choose the problem that is represented in each fiber optic microscope image. Each problem is represented only once.

A. The fiber optic cable is broken or completely blocked.

B. The fiber optic endface is dirty and needs cleaning or repolishing.

C. The fiber optic endface is scratched and needs repolishing or re-termination.

D. The fiber optic cable is cracked across the core and should be replaced.

E. The fiber optic cable is shattered within the cladding and should be re-terminated.

F. The fiber optic endface has a rough surface or has excess epoxy and needs repolishing.

1. _____ Which problem is represented in Microscope Image 1?

2. _____ Which problem is represented in Microscope Image 2?

3. _____ Which problem is represented in Microscope Image 3?

4. _____ Which problem is represented in Microscope Image 4?

5. _____ Which problem is represented in Microscope Image 5?

6. _____ Which problem is represented in Microscope Image 6?

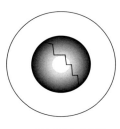

**MICROSCOPE
IMAGE 1**

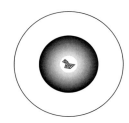

**MICROSCOPE
IMAGE 2**

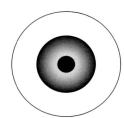

**MICROSCOPE
IMAGE 3**

**MICROSCOPE
IMAGE 4**

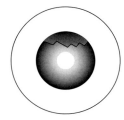

**MICROSCOPE
IMAGE 5**

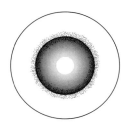

**MICROSCOPE
IMAGE 6**

Name_____ Date_____

_____ 1. Voltage unbalance should not be more than what percent?

_____ 2. What color is standard for grounding conductors?

_____ 3. How many seconds can a momentary power interruption last?

_____ 4. Which harmonic is 180 Hz?

T   F   5. Nonlinear loads are a common source of harmonic problems.

_____ 6. Which test instrument is used to detect noise in an electrical signal?

_____ 7. What colors are standard for neutral conductors?

_____ 8. Is a programmable controller a linear or nonlinear load?

T   F   9. Some motors can continue to operate on only two out of three phases.

_____ 10. Which test instrument is used to test phase unbalance?

_____ 11. Current unbalance should not be more than what percentage?

T   F   12. Improper phase sequence reverses motor rotation.

_____ 13. How many electrical degrees separate balanced phases?

_____ 14. Is an incandescent lamp a linear or nonlinear load?

_____ 15. What type of fluctuation is a voltage decrease of 15% for 32 sec?

T   F   16. High voltage unbalances can be created by small current unbalances.

T   F   17. Harmonics are produced by linear loads.

_____ 18. Which test instrument is designed to measure temperature at a point without contact?

T   F   19. 480 Hz is a triplen harmonic.

_____ 20. What type of fluctuation is a voltage increase of 12% for 78 sec?

21. Explain the difference between linear and nonlinear loads.

22. What effects can harmonics have on loads?

23. How does low power factor affect the cost of operating a power distribution system?

24. Explain the first step when troubleshooting a power quality problem.

25. How can temperature measurements help troubleshoot power quality problems?

Name_____ Date_____

Transformer Output

Three single-phase transformers can be connected in a wye or delta configuration to develop three-phase voltage. A wye configuration is a three-phase transformer connection that has one end of each transformer coil connected together. The remaining end of each coil is connected to the three incoming power lines (primary side) or used to supply power to the loads (secondary side). A delta configuration is a transformer connection that has each of the three transformer coils connected end-to-end to form a closed loop. Each connection point is connected to an incoming power line (primary side) or used to supply power to the loads (secondary side). The voltage output and type available for the loads is determined by whether the transformer is connected in a wye or delta configuration. See Transformer Configurations.

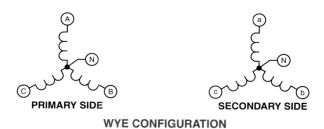

PRIMARY SIDE   **SECONDARY SIDE**

WYE CONFIGURATION

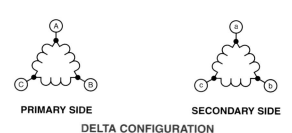

PRIMARY SIDE   **SECONDARY SIDE**

DELTA CONFIGURATION

TRANSFORMER CONFIGURATIONS

The primary side of a three-phase transformer bank may be either a wye or delta configuration, depending on what the utility company has determined to be the best configuration for transmitting and balancing their system. The secondary side of the transformer bank may be either a wye or delta configuration, depending on what is best for the loads. See Wye-to-Wye Step-Down Transformer Bank and Delta-to-Delta Step-Down Transformer Bank.

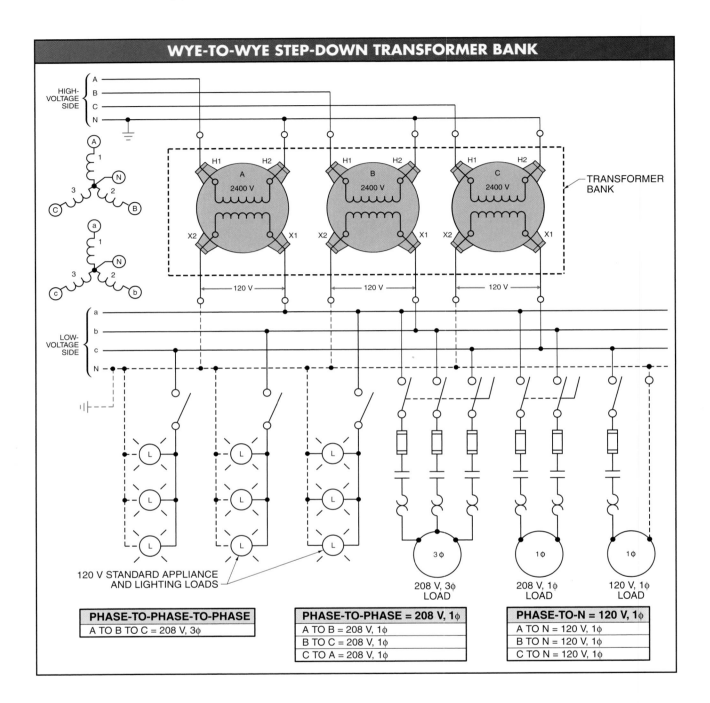

WYE-TO-WYE STEP-DOWN TRANSFORMER BANK

PHASE-TO-PHASE-TO-PHASE
A TO B TO C = 208 V, 3φ

PHASE-TO-PHASE = 208 V, 1φ
A TO B = 208 V, 1φ
B TO C = 208 V, 1φ
C TO A = 208 V, 1φ

PHASE-TO-N = 120 V, 1φ
A TO N = 120 V, 1φ
B TO N = 120 V, 1φ
C TO N = 120 V, 1φ

120 V STANDARD APPLIANCE AND LIGHTING LOADS

208 V, 3φ LOAD

208 V, 1φ LOAD

120 V, 1φ LOAD

A wye-connected secondary is commonly used in schools, stores, and offices and provides three different types of service:

- 208 V, three-phase service for 3φ motor loads and three-phase heaters
- 208 V, single-phase service for 1φ motor loads, single-phase heaters, and single-phase lamps
- 120 V, single-phase service for single-phase lighting and small appliance loads

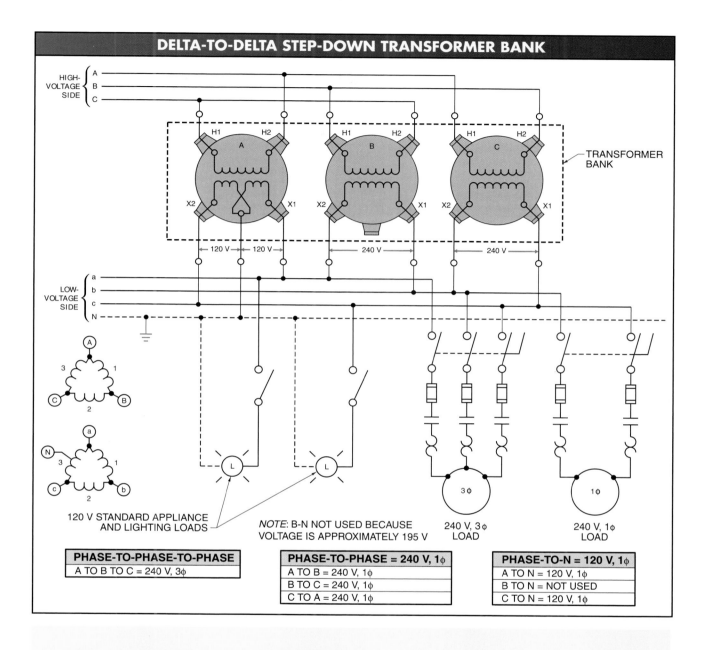

DELTA-TO-DELTA STEP-DOWN TRANSFORMER BANK

120 V STANDARD APPLIANCE AND LIGHTING LOADS

NOTE: B-N NOT USED BECAUSE VOLTAGE IS APPROXIMATELY 195 V

240 V, 3φ LOAD

240 V, 1φ LOAD

PHASE-TO-PHASE-TO-PHASE
A TO B TO C = 240 V, 3φ

PHASE-TO-PHASE = 240 V, 1φ
A TO B = 240 V, 1φ
B TO C = 240 V, 1φ
C TO A = 240 V, 1φ

PHASE-TO-N = 120 V, 1φ
A TO N = 120 V, 1φ
B TO N = NOT USED
C TO N = 120 V, 1φ

A delta-connected secondary provides three different types of service:

- 240 V, three-phase service for 3φ motor loads and three-phase heaters
- 240 V, single-phase service for 1φ motor loads, single-phase heaters, and single-phase lamps
- 120 V, single-phase service for single-phase lighting and small appliance loads

In a basic transformer, if the primary voltage changes (up or down), then the secondary voltage will also change (up or down). However, most power distribution transformers are adjustable because they must maintain the correct voltage output to loads on the secondary side. Taps on the primary side change the ratio of the transformer coils so that the secondary output remains constant even if the primary voltage changes. For example, a common transformer type changes 480 V (primary side) to 120 V (secondary side). The ratio of the transformer is 4:1 (480 V:120 V = 4:1). If the 480 V service sags down to 456 V and the ratio remains constant, then the secondary falls to 114 V (456 V ÷ 4 = 114 V). Instead, the primary side tap is changed so that the ratio becomes 3.8:1, which produces 120 V (456 V ÷ 3.8 = 120 V). Tap adjustments may be manual or automatic. Tap positions are often labeled with the primary side voltage that produces the desired secondary output at that position. See Transformer Taps.

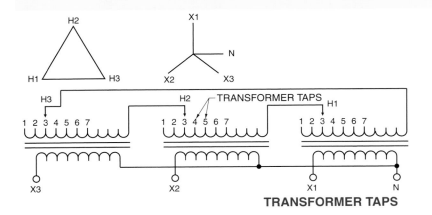

TAP CONNECTIONS		
PRIMARY VOLTAGE	**TAP**	**SECONDARY VOLTAGE**
503	1	208 V (LINE-TO-LINE)
493	2	
480	3	
466	4	
456	5	120 V (LINE-TO-NEUTRAL)
443	6	
433	7	

TRANSFORMER TAPS

Answer the following questions based on the transformer nameplate data. Assume the system loads have a perfect power factor.

1. _____ Is the primary side of this transformer designed for a wye or delta power supply connection?

2. _____ Is the secondary side of this transformer designed for a wye or delta load connection?

3. _____ If this transformer is used to supply 120 V loads, what is the normal primary to secondary voltage ratio?

4. _____ How much power (in VA) can this transformer bank deliver without being overloaded?

5. _____ If the loads are balanced (spread out evenly over the three power lines/individual transformers), how much power (in VA) can each of the three individual transformers deliver without being overloaded?

TRANSFORMER						
DRY TYPE	INDOOR	3φ	60 Hz	CLASS AA		

		JUMPER CONNECTIONS EACH PHASE	
MODEL #	T624A762		
SERIAL #	68A	VOLTS	TAP
kVA	10 150°C RISE	503	1
HV	480 V LINE-TO-LINE	493	2
LV	208 V LINE-TO-LINE	480	3
LV	120 V LINE-TO-NEUTRAL	466	4
WEIGHT	400 LB	456	5
		443	6
H1, H2, H3 = HIGH SIDE		433	7
X1, X2, X3 = LOW SIDE			

HOMEWOOD, IL   MADE IN USA

Answer the following questions based on the transformer nameplate data and the circuit.

6. Set (draw in) the function switch position on Multimeter 1 for taking voltage measurements at the transformer.

7. Connect Multimeter 1 to measure the voltage for any load connected to L2 and L3.

8. Set (draw in) the function switch position on Multimeter 2 for taking voltage measurements at the transformer.

9. Connect Multimeter 2 to measure the voltage for any load connected to L1 and neutral.

10. _____ At the current tap position, what is the primary to secondary voltage ratio of the transformer if the secondary voltage is 120 V?

11. _____ If Multimeter 2 displays 124 V, what is the primary side voltage?

12. _____ If Multimeter 2 displays 124 V, which tap position should the primary tap be moved to?

13. _____ If Multimeter 2 displays 114 V instead, what is the primary side voltage?

14. _____ If Multimeter 2 displays 114 V, which tap position should the primary tap be moved to?

15. _____ If the loads connected to the system have a perfect power factor (100% or 1), what is the maximum current that Multimeter 3 should read?

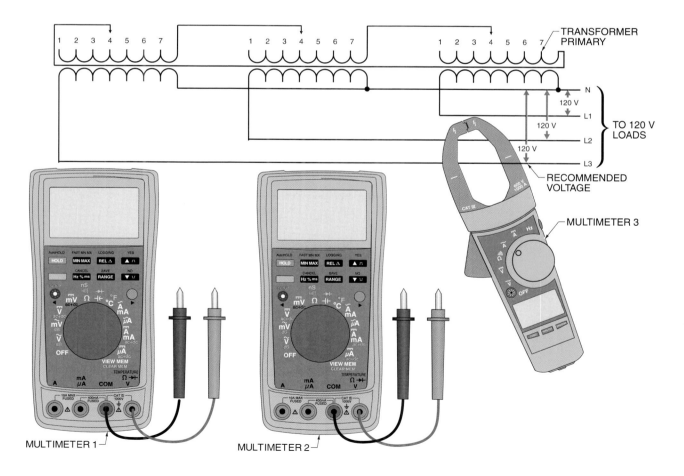

Electric power companies generate and distribute high-voltage, three-phase power. Residential, commercial, and most industrial customers use low-voltage, three-phase (208 V to 480 V) and single-phase (120 V to 277 V) power. Some industrial customers use three-phase voltages higher than 480 V.

Transformers connected in wye or delta configurations step down the high voltage from the transmission lines to the voltage level required by the consumer. Three transformers (except in an open-delta system, which uses only two) are interconnected to deliver single-phase and three-phase voltages to commercial and industrial customers. One transformer is normally used for delivering single-phase voltage to residential customers.

The type, size, and voltage level of a power distribution system are determined by the customer's power needs and the utility company's distribution system. For example, power produced for a large commercial application such as a college, business, or hotel is delivered to the building as high voltage from the utility company. An in-plant substation reduces the high voltage from the utility company for use in building circuits.

A substation is an assembly of equipment installed for switching, changing, or regulating the voltage of electricity. A substation can be a large outdoor utility distribution center or an in-plant distribution center. An in-plant substation contains transformers, switchboards, switchgear, transfer switches, and secondary switches that distribute low voltage levels (208 V to 480 V) to feeder panels, other secondary transformers, busway systems, and branch-circuit panels. See Building Power Distribution.

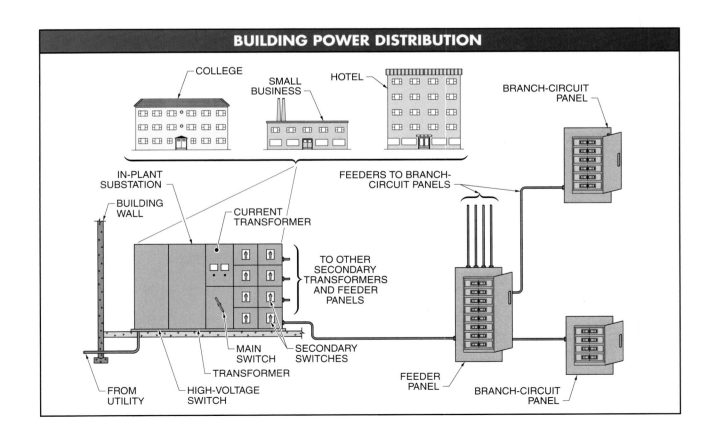

BUILDING POWER DISTRIBUTION

Power quality measurement and troubleshooting activities vary with the type of distribution system—whether it is a residential, commercial, or industrial application. When troubleshooting power quality problems, measurements such as voltage, current, power, or harmonics are taken throughout the building distribution system. An understanding of the different types of power distribution systems, power quality problems, and test instruments is required when troubleshooting a problem. A Power Quality Troubleshooting Checklist can be used to help diagnose problems or identify potential problem areas. See Appendix.

Conductor Color Coding

The type of system (wye or delta) and voltage levels must be known when taking electrical measurements or troubleshooting an electrical circuit. Standard conductor color codes help determine the type of electrical circuit. Conductor color coding is also helpful when troubleshooting, so it is even used sometimes in applications that do not require every conductor to be color-coded. Always use standard colors if possible.

Conductors are covered with an insulating material that is available in different colors. The advantage of using different colors on conductors is that the color shows the function of each conductor. Some colors have a standard meaning. For example, the color green always indicates a conductor used for grounding. Other colors may have more than one meaning depending on the circuit. For example, a red conductor may be used to indicate a hot wire in a 230 V circuit or switched wire in a 115 V circuit.

Green, or green with a yellow stripe, is the standard color for a grounding conductor. Green is used to indicate a grounding conductor regardless of the voltage level (such as 115 V, 230 V, or 460 V) or circuit (single-phase or three-phase). A grounding conductor is a conductor that normally carries current only during a fault (short circuit). Grounding conductors may also be bare in some applications.

The colors white or natural gray are used for the neutral conductor. A neutral conductor is a current-carrying conductor that is intentionally grounded. Neutral conductors carry current from a load (such as a lamp, heating element, or motor) back to the power source. Neutral conductors are connected directly to loads and never connected through fuses, circuit breakers, or switches.

An ungrounded (hot) conductor is a current-carrying conductor that is connected to loads through fuses, circuit breakers, and switches. Ungrounded conductors can be any color other than white, natural gray, green, or green with a yellow stripe. Black, red, blue, orange, and yellow are usually used for ungrounded conductors, with black being the most common. The exact color used to indicate different hot conductors may vary. For example, the colors used to identify A (L1), B (L2), and C (L3) in a three-phase system depend on the configuration of the system. The exception to this is listed in NEC® Article 110.15, which states that in a 4-wire delta-connected secondary system, the higher voltage phase must be colored orange (or clearly marked) because it is too high for low-voltage (115 V) single-phase power and too low for high-voltage (230 V) single-phase power. See Circuit Conductor Color Coding.

Receptacles (outlets) have different configurations to prevent the connection of equipment to the wrong source of power (voltage type and level). The receptacle must be wired to the correct transformer/service type in order for the receptacle to deliver the proper voltage.

CIRCUIT CONDUCTOR COLOR CODING

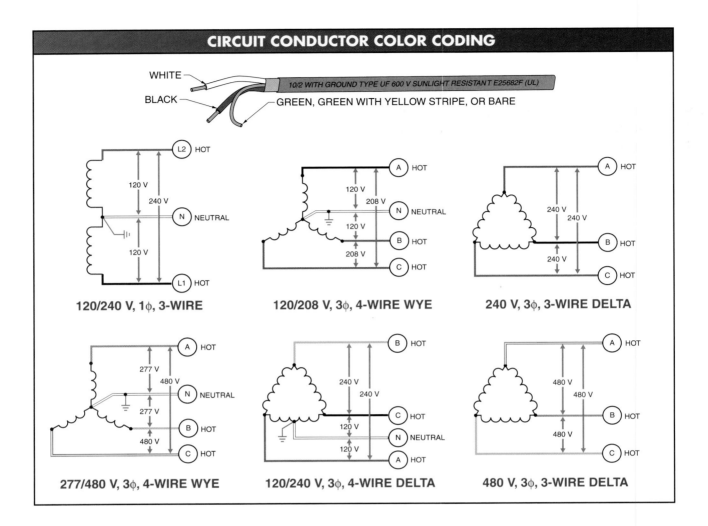

120/240 V, 1φ, 3-WIRE

120/208 V, 3φ, 4-WIRE WYE

240 V, 3φ, 3-WIRE DELTA

277/480 V, 3φ, 4-WIRE WYE

120/240 V, 3φ, 4-WIRE DELTA

480 V, 3φ, 3-WIRE DELTA

1. What are the voltage, phase type, and number of wires for this service?

2. _____ What are the two possible colors of Conductor 1?

3. _____ What is the color of Conductor 2?

4. _____ What is the color of Conductor 3?

5. _____ What are the two possible colors of Conductor 4?

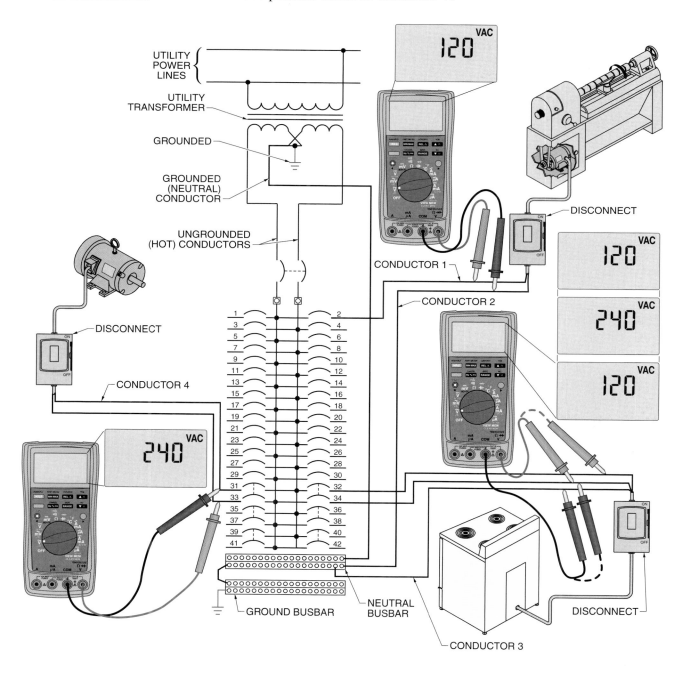

6. What are the voltage, phase type, and number of wires for this service?

7. _____ What is the color of Conductor 1?

8. _____ What is the color of Conductor 2?

9. _____ What is the color of Conductor 3?

10. _____ What is the color of Conductor 4?

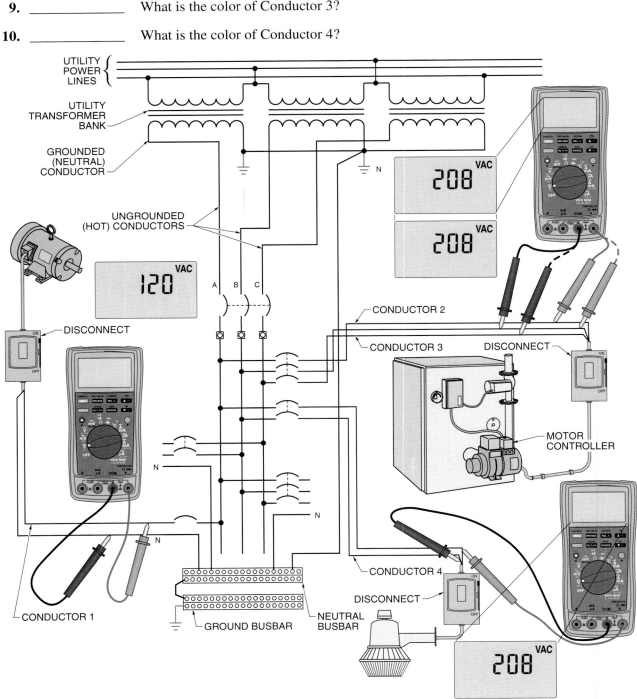

11. What are the voltage, phase type, and number of wires for this service?

12. _____ What is the color of Conductor 1?

13. _____ What is the color of Conductor 2?

14. _____ What is the color of Conductor 3?

15. _____ What is the color of Conductor 4?

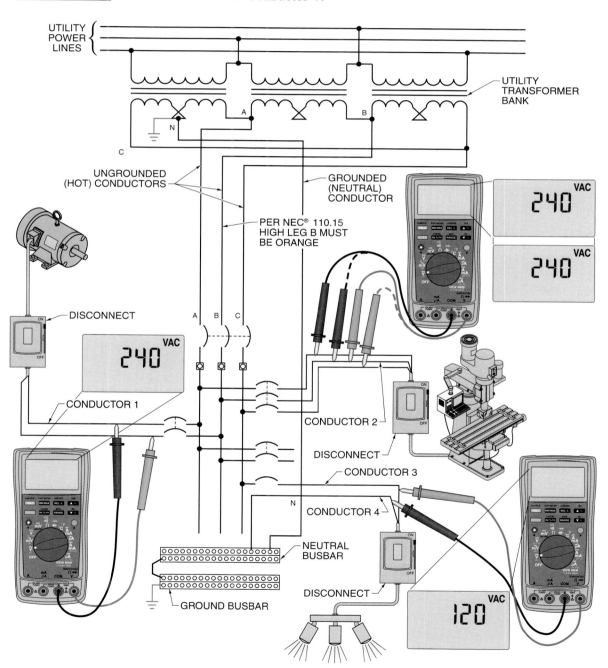

16. What are the voltage, phase type, and number of wires for this service?

17. _____ What is the color of Conductor 1?

18. _____ What is the color of Conductor 2?

19. _____ What is the color of Conductor 3?

20. _____ What is the color of Conductor 4?

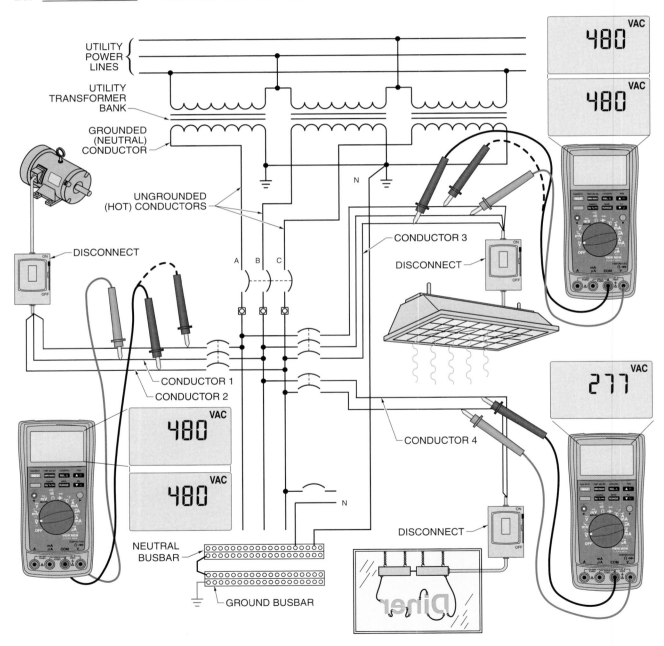

21. _____ What is the voltage and phase type for the commercial receptacle?

22. _____ What is the voltage and phase type for the residential receptacle?

23. _____ What is the voltage and phase type for the industrial receptacle?

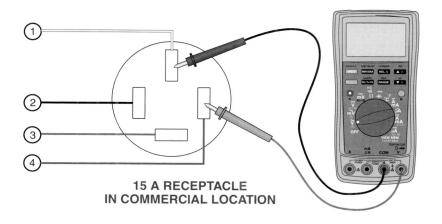

**15 A RECEPTACLE
IN COMMERCIAL LOCATION**

VOLTAGE MEASUREMENTS
1 TO 2 = 120 V
1 TO 4 = 120 V
1 TO 3 = 120 V
2 TO 3 = 208 V
2 TO 4 = 208 V
4 TO 3 = 208 V

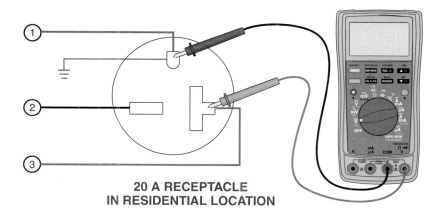

**20 A RECEPTACLE
IN RESIDENTIAL LOCATION**

VOLTAGE MEASUREMENTS
1 TO 2 = 120 V
1 TO 3 = 120 V
2 TO 3 = 240 V

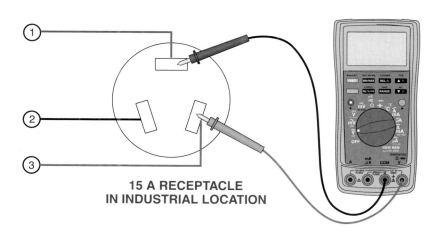

**15 A RECEPTACLE
IN INDUSTRIAL LOCATION**

VOLTAGE MEASUREMENTS
1 TO 2 = 240 V
1 TO 3 = 240 V
2 TO 3 = 240 V

Transformers are used to deliver power to a set number of loads. As loads in a system are switched ON and OFF, the power delivered by the transformer changes. For example, at night, the power output required from a transformer in an office building may be low. During business hours, the power output required from the transformer may be high. In order for a transformer to deliver enough power at all times, the transformer must be able to deliver enough power during peak load demands. Peak load is the maximum output requirement of a transformer. See Transformer Load Cycle.

A transformer is overloaded when it is required to deliver more power than the listed rating. A transformer is usually not damaged when overloaded for a short time because it takes some time for the extra load to raise the transformer's internal temperature to damaging levels. A transformer may be damaged if overloaded for a long time because of improper cooling or power quality problems that cause additional heat buildup. Nonlinear loads cause additional heat from harmonics, which increase at higher frequencies. Nonlinear loads also produce additional heat from the skin effect caused by the higher frequency harmonics.

Transformers that deliver power to many nonlinear loads can be overloaded because of the increased harmonics. Older transformers that delivered enough power when most loads were linear may not be able to deliver enough power when nonlinear loads are added. This occurs even if the total current draw is the same or less than it was when the transformer was delivering power to only linear loads. This is a common problem for transformers delivering power to office buildings, schools, and libraries because large numbers of computers have been added in recent years.

Transformer manufacturers list the length of time a transformer may safely be overloaded at a given peak level. For example, a transformer that is overloaded 15 times its rated current has a permissible overload time of 5.5 sec. The rated permissible overload is based on 60 Hz linear loads that decrease transformer heating as the loads are removed. Additional heating occurs when different load types are connected to a transformer. See Transformer Overloading.

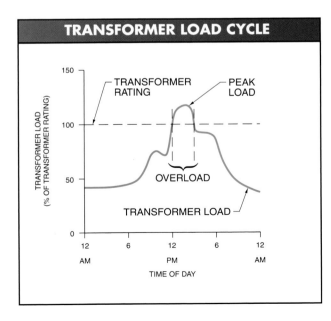

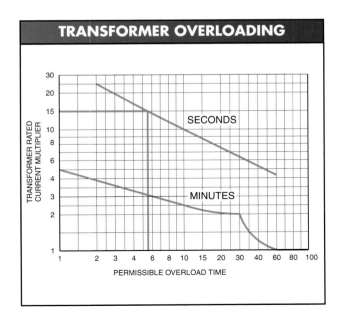

A K-rated transformer is a transformer designed to handle the extra heating effects caused by harmonic currents. A K factor represents the amount of harmonics produced by nonlinear loads connected to power lines, and can be measured with a power quality meter. Larger K factors indicate a greater amount of harmonics present. K-rated transformers have a K-rating such as K-4, K-13, or K-20. Higher K-ratings indicate greater heat dissipation capabilities. See K-Rated Transformers.

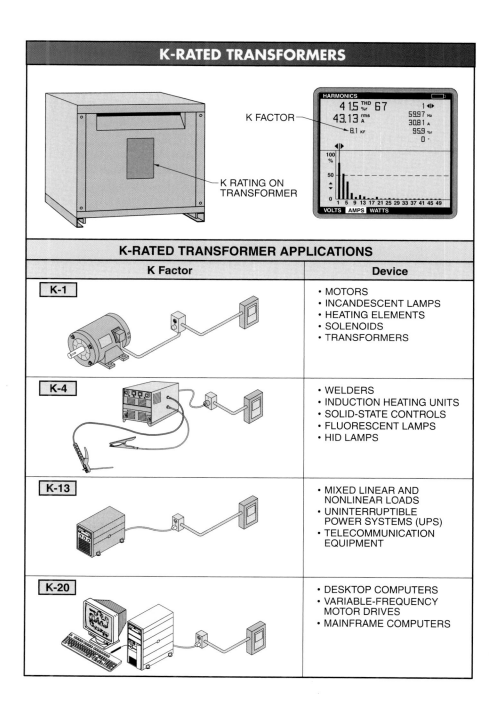

K-RATED TRANSFORMERS

K FACTOR

K RATING ON TRANSFORMER

K-RATED TRANSFORMER APPLICATIONS

K Factor	Device
K-1	• MOTORS • INCANDESCENT LAMPS • HEATING ELEMENTS • SOLENOIDS • TRANSFORMERS
K-4	• WELDERS • INDUCTION HEATING UNITS • SOLID-STATE CONTROLS • FLUORESCENT LAMPS • HID LAMPS
K-13	• MIXED LINEAR AND NONLINEAR LOADS • UNINTERRUPTIBLE POWER SYSTEMS (UPS) • TELECOMMUNICATION EQUIPMENT
K-20	• DESKTOP COMPUTERS • VARIABLE-FREQUENCY MOTOR DRIVES • MAINFRAME COMPUTERS

Linear loads without any harmonics, such as motors, incandescent lamps, and heating elements, have a K factor of 1 (K-1). Nonlinear loads with some harmonics, such as welding machines, induction heating units, and solid-state controls, have a K factor of 4 (K-4). Branch circuits that include a mix of linear and nonlinear loads, such as uninterruptible power systems and telecommunication equipment, have a K factor of 13 (K-13). Circuits that include mostly nonlinear loads, such as desktop computers and variable-frequency motor drives, have a K factor of 20 (K-20). The heating effect caused by the harmonics of a K-4 load is four times that of K-1. The heating effect of a K-20 load is 20 times that of K-1.

K-rated transformers have a larger neutral conductor and special windings to reduce the effects of harmonics and the extra heat they produce. If a transformer does not have a K-rating listed, it is assumed to be a standard transformer with a K-rating of K-1. To determine the required transformer K-rating, a K factor measurement is taken with a power quality meter when the system is fully loaded. The K factor measurement will probably not equal the exact K-rating of available transformers, so the K factor should be rounded up to the next available rating to ensure proper transformer operation under full load.

1. _____   What is the lowest recorded voltage measurement?

2. _____   What is the highest recorded voltage measurement?

3. _____   What is the lowest recorded current measurement?

4. _____   What is the highest recorded current measurement?

5. What caused the voltage to dip to the lowest recorded value?

6. What indicates that the transformer is under the greatest load?

7. _____   If the transformer is overloaded to twice its rating, how long can it be safely overloaded?

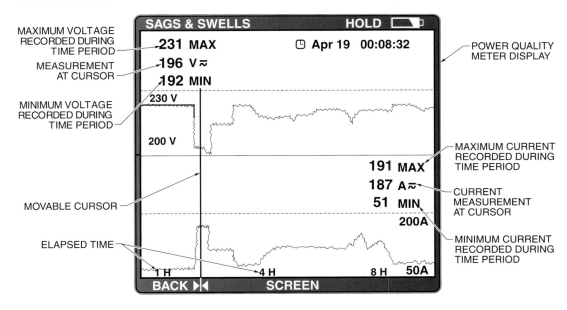

8. _____ What is the voltage at the cursor?

9. _____ What is the current at the cursor?

10. _____ If the circuit loads are rated to operate at 230 VAC, what is the maximum percent voltage drop during the recording time?

11. _____ If the service panel is rated at 200 A, what is the maximum percent of current drawn?

12. _____ If the maximum current was recorded at 12 PM, does this coincide with a typical transformer load cycle?

13. _____ If the loads are mostly computers, printers, and copy machines, which K rating should this transformer have?

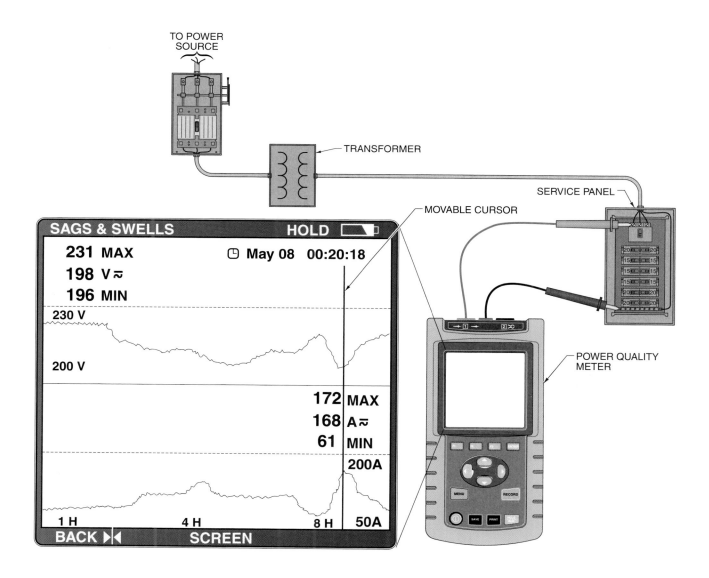

14. _____ Is a high current draw causing the voltage to dip?

15. _____ Is the voltage dip caused by the loads connected to the transformer or a problem upstream from the transformer?

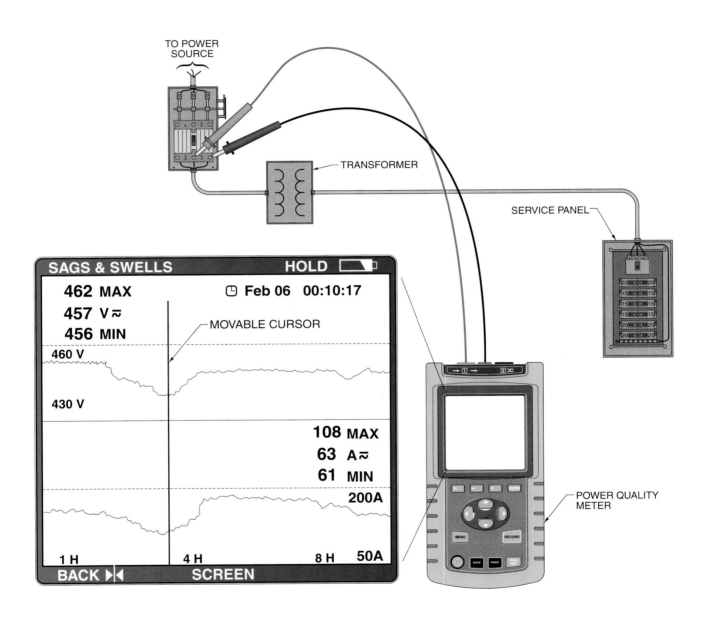

Application 4-4

The electrical service to most residences is 120/240 V, single-phase. The lower voltage line (120 V) is used for general-purpose receptacles and lighting, and is obtained by connecting loads between the neutral conductor and either hot wire (L1 or L2). The higher voltage line (240 V) is used for heating, cooling, and cooking, and is obtained by connecting loads between the two hot wires (L1 and L2).

The current is 180° out of phase between each hot conductor. If they are equal in magnitude, they cancel each other in the neutral wire. For example, if a 10 A linear load is connected to L1 and neutral and a 10 A linear load is connected to L2 and neutral, there is no current flowing in the neutral (10 A – 10 A = 0 A). See Same Current Draw.

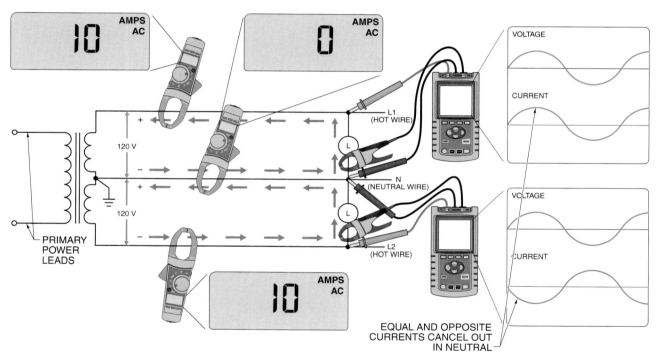

SAME CURRENT DRAW

If linear loads connected between L1 and neutral and L2 and neutral do not have the same current draw, the difference (unbalance) in current draw flows through the neutral. For example, if a 20 A linear load is connected to L1 and neutral and a 5 A linear load is connected to L2 and neutral, there is 15 A of current in the neutral (20 A – 5 A = 15 A). Because the neutral carries only the unbalance current, current in the neutral never exceeds the highest current level in either hot line (L1 or L2) when linear loads are connected to the transformer. See Different Current Draw.

DIFFERENT CURRENT DRAW

Although an ammeter can measure the current cancellation in the neutral, the ammeter cannot show that the cancellation takes place due to the 180° phase shift. A power quality meter can be used to measure the current in the conductors and show the phase shift.

The electrical service to most commercial buildings is a wye-connected, three-phase system. This system is used because it can provide a large amount of balanced single-phase power in addition to three-phase power. In a three-phase, four-wire system supplying power to linear loads, the fundamental 60 Hz currents cancel in the neutral conductor, so one common neutral is usually used.

However, when single-phase nonlinear loads are connected to a three-phase, 4-wire system, neutral current can exceed an individual phase current. The neutral current can be between 125% and 225% of the highest phase current. The nonlinear loads produce triplen harmonics, which are odd-numbered multiples of the third harmonic (such as 3rd, 9th, 15th, and so on). Triplen harmonics do not cancel, but add together in the neutral conductor. See Harmonics from Nonlinear Loads.

The third harmonic current is usually responsible for most of the neutral current because it is typically the harmonic with the highest current value. High neutral current is dangerous because it causes overheating in the neutral conductor. Because there is no circuit breaker to limit current, overheating of the neutral can become a fire hazard. Excessive current in the neutral conductor can also cause higher than normal voltage drops between the neutral and ground at 120 V outlets.

When nonlinear loads are connected to power lines, any harmonics present cause the 60 Hz fundamental frequency to become distorted. This distortion causes inaccurate measurements in test instruments that are not rated "true-rms." To prevent inaccurate measurements, only a true-rms multimeter or power quality meter should be used when taking voltage or current measurements.

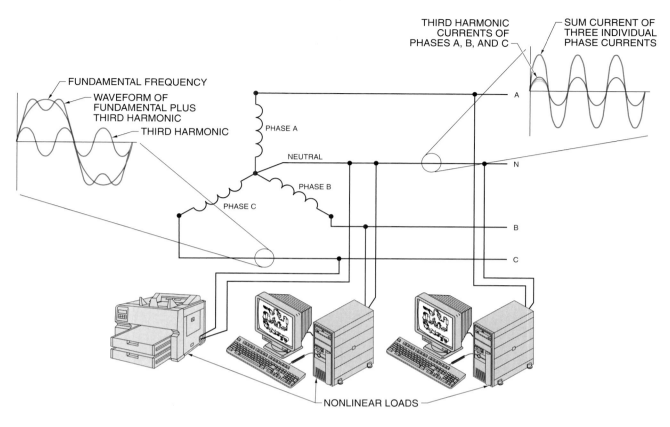

HARMONICS FROM NONLINEAR LOADS

A power quality meter can also display the percent of harmonic current on the power lines. The power quality meter usually displays the odd-number harmonics (1, 3, 5, and so on) since it is the odd-numbered harmonics that cause system and load problems, such as overheating. See Harmonics Display.

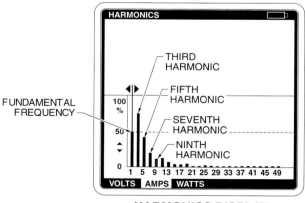

HARMONICS DISPLAY

1. _____ Which measurement shows that the loads connected to the system are all linear loads?

2. _____ Which measurement shows that the loads connected to the system are more linear than nonlinear?

3. _____ Which measurement shows that many of the loads must be nonlinear?

4. _____ What frequency would the multimeter display for Measurement 1?

5. _____ What frequency would the multimeter display for Measurement 2?

6. _____ What frequency would the multimeter display for Measurement 3?

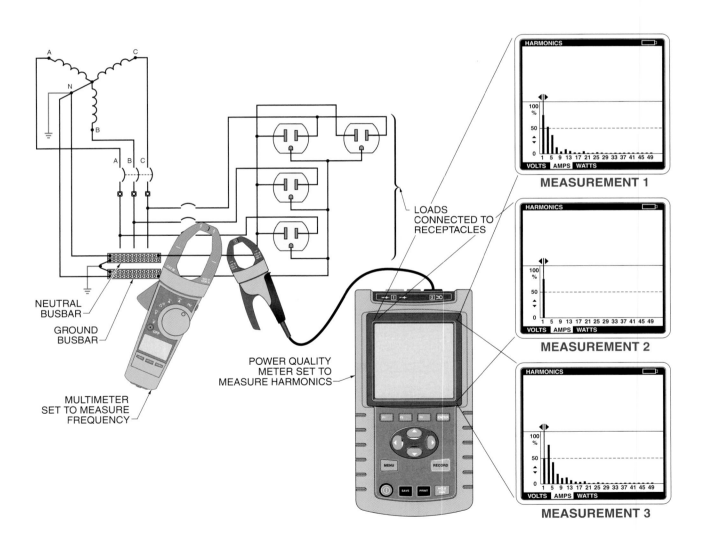

NEUTRAL BUSBAR

GROUND BUSBAR

MULTIMETER SET TO MEASURE FREQUENCY

POWER QUALITY METER SET TO MEASURE HARMONICS

LOADS CONNECTED TO RECEPTACLES

MEASUREMENT 1

MEASUREMENT 2

MEASUREMENT 3

Application 4-5   *Voltage Changes*

It is normal for AC voltage to vary slightly, and fluctuations within +5% to –10% of the voltage rating of loads are usually not a problem. However, more serious voltage changes, such as power interruptions, voltage sags, voltage swells, undervoltages, and overvoltages, are a problem. Voltmeters with a recording mode (MIN MAX) can be used to identify voltage changes within the meter's recording specifications. For example, a meter with a 100 ms (0.1 sec) capture time can record any voltage changes that occur for at least 6 cycles (0.1 sec × 60 cycles per sec = 6 cycles). This means the meter can capture most voltage changes, but not transient voltages, which typically only last a fraction of a cycle. See Voltage Changes.

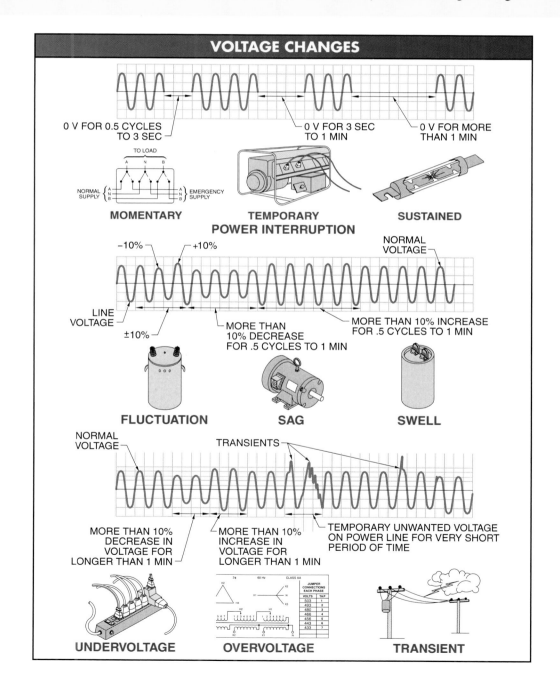

VOLTAGE CHANGES

It is reported that computers connected to Circuit 4 have been automatically rebooting. To diagnose the problem, voltage measurements are taken over time at different points in the power distribution system. Answer the following questions based on the voltage measurement results.

1. Why are voltage measurements taken at the receptacles on Circuit 4?

2. _____ Is there a problem at Circuit 4 that could cause computers to automatically reset?

3. Why are voltage measurements taken at the main circuit breaker in Panelboard 2?

4. _____ Is there a problem at Circuit 4 only?

5. Why are voltage measurements taken at the main circuit breaker in Panelboard 1?

6. _____ Is there a problem throughout the entire in-plant power distribution system?

7. Why are voltage measurements taken at the main power feed into the switchboard?

8. Where is the most likely problem and a place where additional voltage measurements need to be taken?

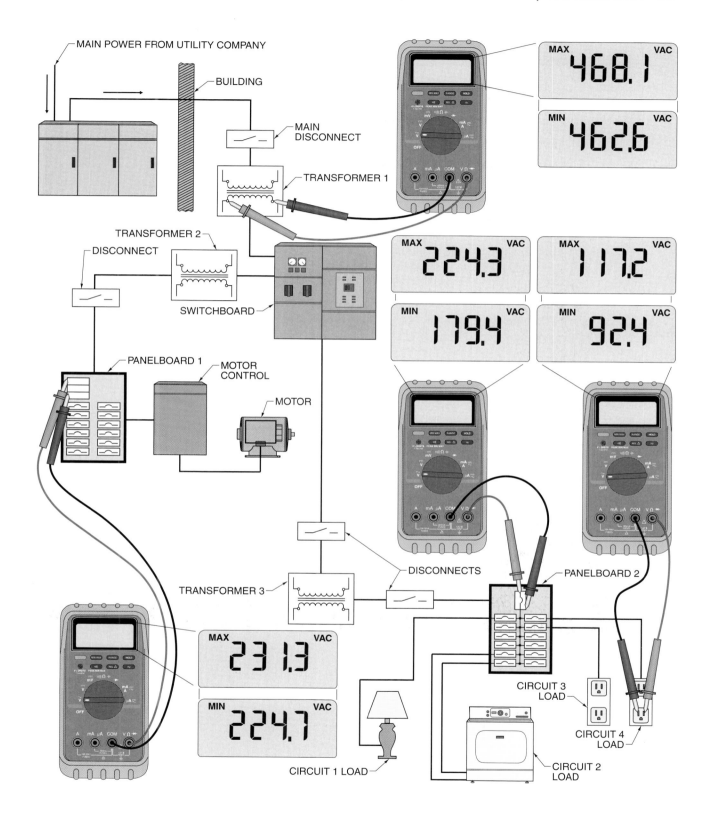

Normal AC voltage and current values generate a waveform in the shape of the sine function. Certain loads may distort the sinusoidal (sine function) waveform. Crest factor is the measure of the amount of distortion in nonsinusoidal waveforms.

AC voltage values are stated and measured as peak, peak-to-peak, average, or rms values. The peak voltage value (V_{max}) of a sine wave is the maximum value of either the positive or negative alternation. The positive and negative alternation values are equal in a sine wave.

The peak-to-peak voltage value (V_{p-p}) is the value measured from the maximum positive alternation to the maximum negative alternation. Some voltmeters are designed to measure peak-to-peak voltage. See Peak-to-Peak Voltage.

The average voltage value (V_{avg}) of a sine wave is the mathematical mean of all instantaneous voltage values in the sine wave. The average voltage value is equal to 0.637 times the peak value of a standard sine wave. See Average Voltage.

The root-mean-square voltage value (V_{rms}), or effective value, of a sine wave is the voltage value that produces the same amount of heat in a pure resistive circuit as DC of the same value. The rms value is equal to 0.707 times the peak value in a sine wave. See rms Voltage.

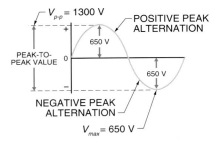

PEAK-TO-PEAK VOLTAGE

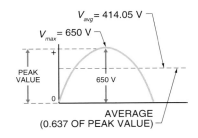

AVERAGE VOLTAGE

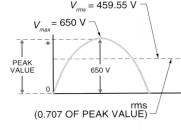

rms VOLTAGE

Some voltmeters and ammeters, especially older models, measure the peak voltage or current value and then display an equivalent rms value that is calculated by multiplying the peak value by 0.707. For example, a voltmeter connected to a 120 VAC receptacle may measure the circuit's peak voltage (170 VAC) and display the equivalent rms value (170 VAC × 0.707 = 120 VAC). This value is accurate if the waveform is a perfectly sinusoidal waveform.

Linear loads do not distort the sine wave. For example, when a magnetic motor starter is used to control a motor, the motor and motor starter load will not distort the AC voltage sine wave or AC current sine wave. See Linear Load Crest Factors.

A nonsinusoidal waveform has a distorted appearance when compared with a pure sine waveform. Nonsinusoidal waveforms are present in equipment such as variable-speed motor drives, light dimmers, computers, and most circuits that use solid-state electronics. These types of loads are called nonlinear loads.

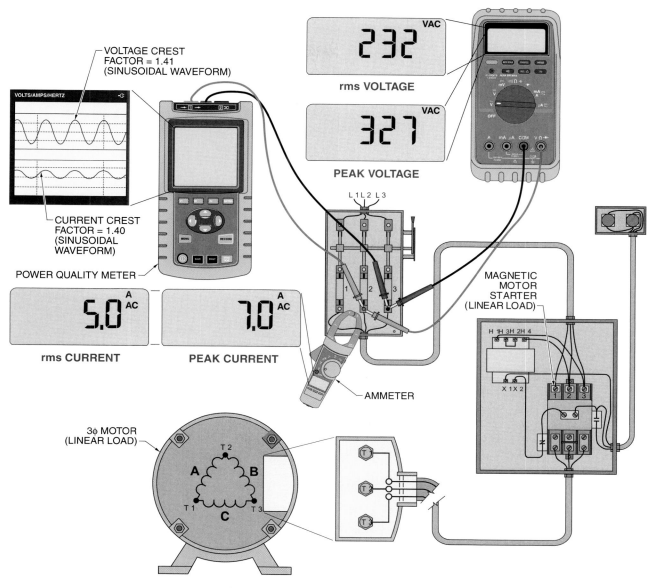

VOLTAGE CREST
FACTOR = 1.41
(SINUSOIDAL WAVEFORM)

VOLTS/AMPS/HERTZ

CURRENT CREST
FACTOR = 1.40
(SINUSOIDAL
WAVEFORM)

POWER QUALITY METER

232 VAC

rms VOLTAGE

327 VAC

PEAK VOLTAGE

L1 L2 L3

MAGNETIC
MOTOR
STARTER
(LINEAR LOAD)

5.0 A AC

rms CURRENT

7.0 A AC

PEAK CURRENT

AMMETER

3φ MOTOR
(LINEAR LOAD)

LINEAR LOAD CREST FACTORS

If the AC waveform is distorted, a calculated rms value is not correct. Voltage measurement errors can range from about 1% to more than 50%. This is a problem as systems include more nonlinear loads. Meters that are rated "true-rms" measure rms values directly, so they are the most accurate meters for any waveform shape, including nonsinusoidal. Only true-rms test instruments should be used in any circuit that includes a nonlinear load. For example, when a solid-state motor starter is used to control a motor, the motor starter load will distort the AC current waveform. See Nonlinear Load Crest Factors.

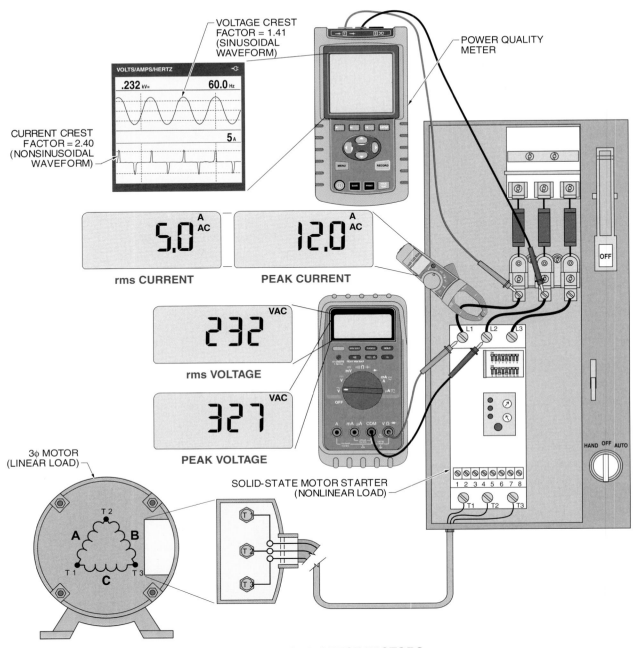

NONLINEAR LOAD CREST FACTORS

Crest factor is the ratio of the peak voltage value to the rms voltage value (or the peak current value to the rms current value). A pure sinusoidal waveform has a crest factor of 1.41. Distortion of the waveform changes the crest factor. A greater change in crest factor from 1.41 indicates a more distorted waveform, which reduces the accuracy of calculated rms values. See High Crest Factors.

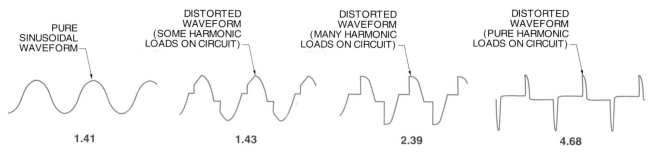

PURE SINUSOIDAL WAVEFORM	DISTORTED WAVEFORM (SOME HARMONIC LOADS ON CIRCUIT)	DISTORTED WAVEFORM (MANY HARMONIC LOADS ON CIRCUIT)	DISTORTED WAVEFORM (PURE HARMONIC LOADS ON CIRCUIT)
1.41	1.43	2.39	4.68

HIGH CREST FACTORS

Crest factor is calculated by applying the following formula:

$$CF = \frac{V_p}{V_{rms}}$$

where

CF = crest factor

V_p = peak voltage (in V)

V_{rms} = rms voltage (in V)

For example, what is the crest factor of a voltage waveform with peak voltage of 294 V and true rms voltage of 208 V?

$$CF = \frac{V_p}{V_{rms}}$$

$$CF = \frac{294}{208}$$

$$CF = \mathbf{1.41}$$

The crest factor is 1.41, so the voltage waveform must be sinusoidal. However, the same power circuit has a peak current of 12.7 A and true rms current of 5.3 A, resulting in a crest factor of 2.4 (12.7 A ÷ 5.3 A = 2.4). This indicates that many nonlinear loads in this circuit are distorting the current waveform.

1. _____ What is the crest factor of the voltage?

2. _____ What is the crest factor of the current?

3. Draw the expected voltage waveform for the circuit in the oscilloscope screen.

4. Draw the expected current waveform for the circuit in the oscilloscope screen.

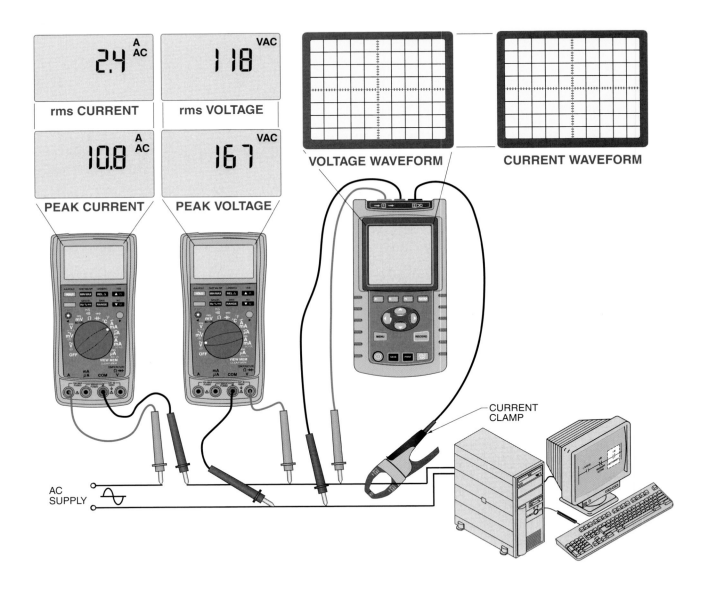

5. _____ What is the crest factor of the voltage?

6. _____ What is the crest factor of the current?

7. Draw the expected voltage waveform for the circuit in the oscilloscope screen.

8. Draw the expected current waveform for the circuit in the oscilloscope screen.

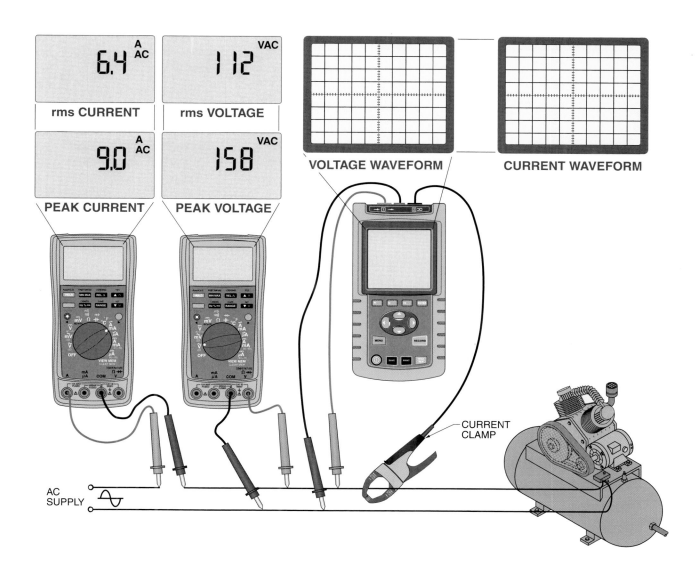

9. _____ What is the crest factor of the voltage?

10. _____ What is the crest factor of the current?

11. Draw the expected voltage waveform for the circuit in the oscilloscope screen.

12. Draw the expected current waveform for the circuit in the oscilloscope screen.

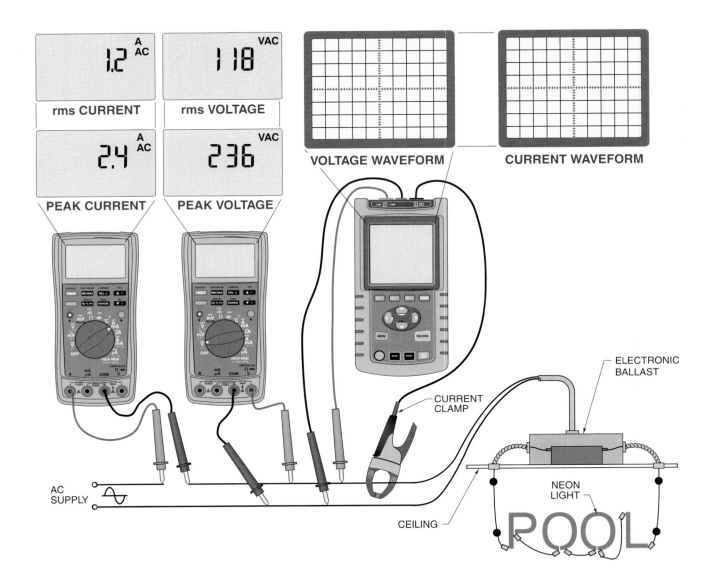

13. _____ What is the crest factor of the voltage?

14. _____ What is the crest factor of the current?

15. Draw the expected voltage waveform for the circuit in the oscilloscope screen.

16. Draw the expected current waveform for the circuit in the oscilloscope screen.

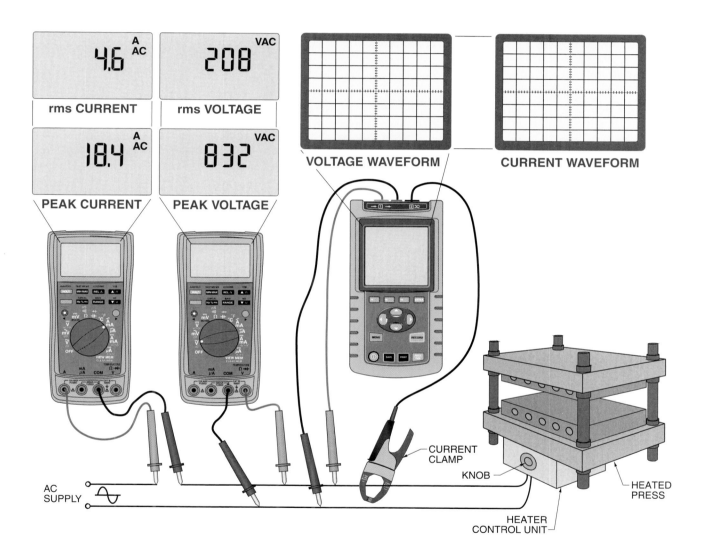

Low Crest Factor

Nonlinear loads causing higher than normal current crest factors (greater than 1.41) can also produce lower than normal voltage crest factors (less than 1.41). When the high current is drawn at the voltage peak, it causes the voltage peak to be pulled down. This pulling down of voltage peaks is called "flat-topping." See Flat-Topped Voltage Waveform and Distorted Current Waveform.

Severe flat-topping can cause the normal voltage sine wave to appear like a square wave. This shape results in the peak and rms values being closer together. When crest factor is calculated, the value is less than 1.41. (The minimum possible crest factor is 1.0, which describes a perfect square wave.)

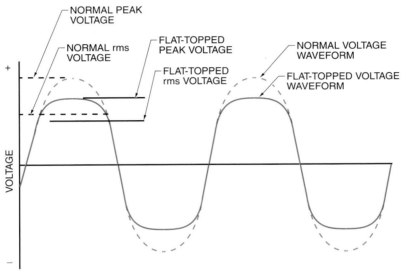

FLAT-TOPPED VOLTAGE WAVEFORM

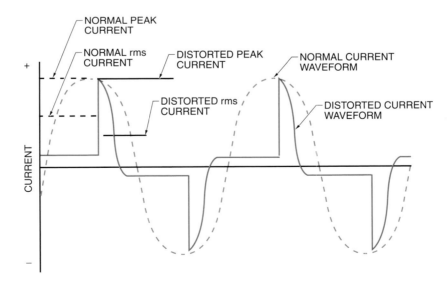

DISTORTED CURRENT WAVEFORM

A voltage flat-topping problem can be detected by calculating crest factor from peak and rms voltages. By calculating current crest factor as well, the problem can be attributed to distortion in the current waveform rather than some other problem.

1. _____ What is the voltage crest factor?

2. _____ What is the current crest factor?

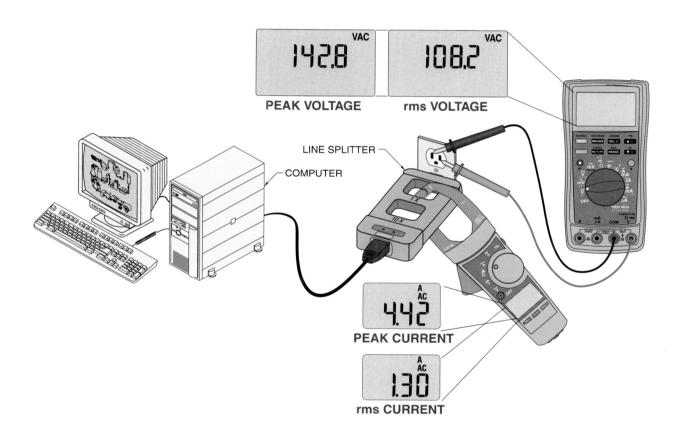

3. _____ What is the voltage crest factor?

4. _____ What is the current crest factor?

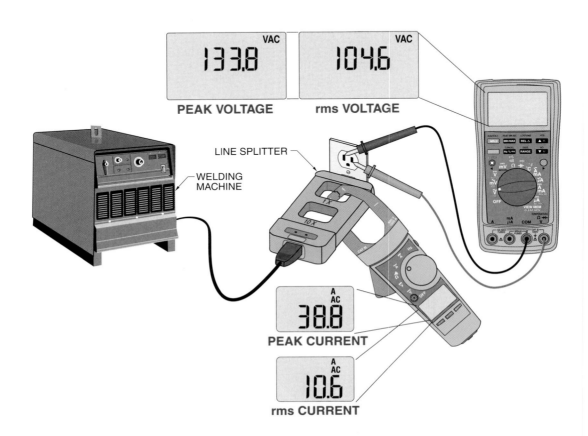

5. _____ What is the voltage crest factor?

6. _____ What is the current crest factor?

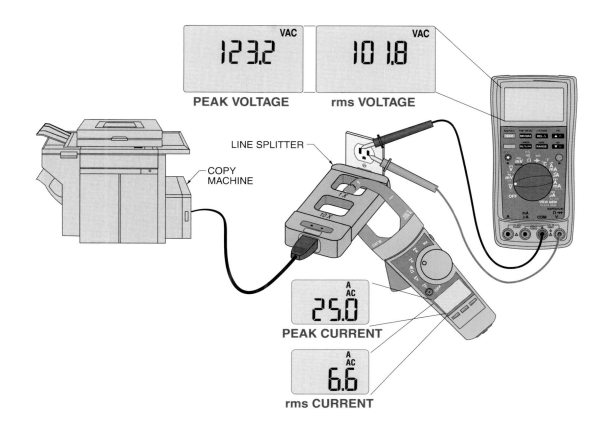

Application 4-8 — *Voltage and Current Unbalance*

Multimeters can be used to measure voltage and current unbalance in a three-phase power system by taking readings at each line. Three separate measurements must be taken and recorded. Unbalance is determined by calculating the percent difference between the largest deviation and the average value.

A three-phase power quality meter can also be used to measure the voltage and current unbalance. This is the quickest method because the meter is connected to each power line with voltage leads and current clamps and takes readings simultaneously. The meter also makes the calculations required to determine unbalance and displays the result directly.

1. _____ What is the percentage of voltage unbalance at the motor?

2. _____ Is the voltage unbalance within acceptable limits?

3. _____ What is the percentage of current unbalance at the motor?

4. _____ Is the current unbalance within acceptable limits?

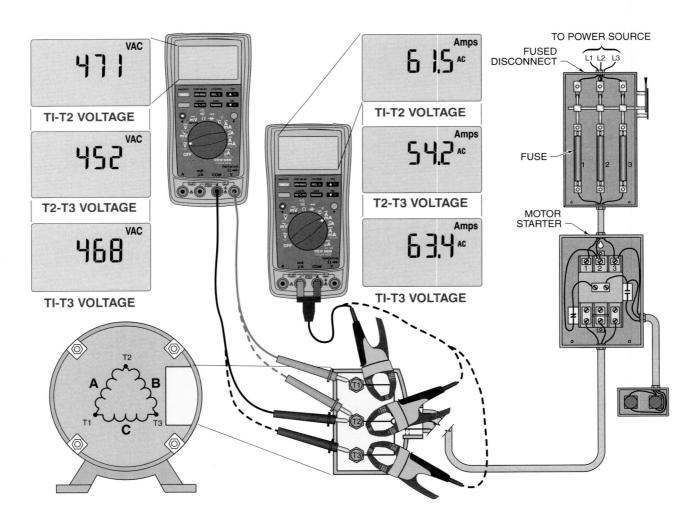

471 VAC	**61.5** Amps AC
TI-T2 VOLTAGE	TI-T2 VOLTAGE
452 VAC	**54.2** Amps AC
T2-T3 VOLTAGE	T2-T3 VOLTAGE
468 VAC	**63.4** Amps AC
TI-T3 VOLTAGE	TI-T3 VOLTAGE

5. _____ What is the percentage of voltage unbalance at the motor?

6. _____ What is the percentage of current unbalance at the motor?

7. _____ Based on the motor's nameplate information, are the current measurements within acceptable limits?

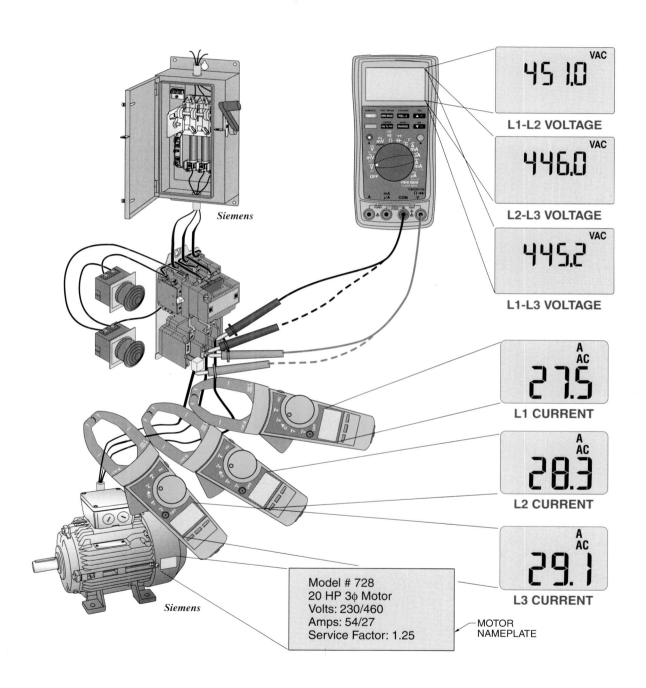

L1-L2 VOLTAGE — 451.0 VAC

L2-L3 VOLTAGE — 446.0 VAC

L1-L3 VOLTAGE — 445.2 VAC

L1 CURRENT — 27.5 A AC

L2 CURRENT — 28.3 A AC

L3 CURRENT — 29.1 A AC

Siemens

Model # 728
20 HP 3φ Motor
Volts: 230/460
Amps: 54/27
Service Factor: 1.25

MOTOR NAMEPLATE

8. _____ What is the percentage of voltage unbalance at the motor?

9. _____ What is the percentage of current unbalance at the motor?

10. _____ Are the current measurements within an acceptable range for a 20 HP motor with a nameplate current that could not be read? (The motor might be in a remote location, or the nameplate damaged or missing.)

11. _____ Do the measurements indicate that the voltage waveform is sinusoidal or distorted?

12. _____ Do the measurements indicate that the current waveform is sinusoidal or distorted?

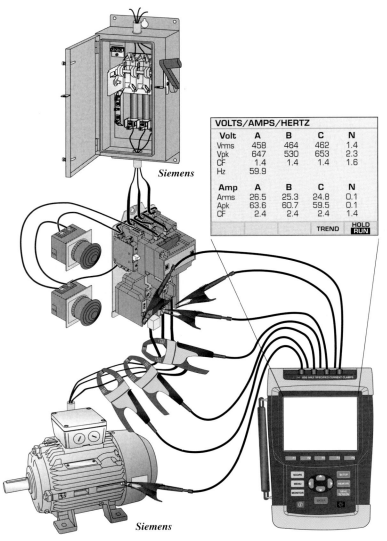

VOLTS/AMPS/HERTZ

Volt	A	B	C	N
Vrms	458	464	462	1.4
Vpk	647	530	653	2.3
CF	1.4	1.4	1.4	1.6
Hz	59.9			

Amp	A	B	C	N
Arms	26.5	25.3	24.8	0.1
Apk	63.6	60.7	59.5	0.1
CF	2.4	2.4	2.4	1.4

TREND   HOLD RUN

Siemens

Motor Rating*	Current†			
	208 V	230 V	460 V	575 V
1/4	1.11	0.96	0.48	0.38
1/3	1.34	1.18	0.59	0.47
1/2	2.2	2.0	1.0	0.8
3/4	3.1	2.8	1.4	1.1
1	4.0	3.6	1.8	1.4
1 1/2	5.7	5.2	2.6	2.1
2	7.5	6.8	3.4	2.7
3	10.6	9.6	4.8	3.9
5	16.7	15.2	7.6	6.1
7 1/2	24.0	22.0	11.0	9.0
10	31.0	28.0	14.0	11.0
15	46.0	42.0	21.0	17.0
20	59	54	27	22
25	75	68	34	27
30	88	80	40	32
40	114	104	52	41
50	143	130	65	52
60	169	154	77	62
75	211	192	96	77
100	273	248	124	99
125	343	312	156	125
150	396	360	180	144
200	—	480	240	192
250	—	602	301	242
300	—	—	362	288
350	—	—	413	337
400	—	—	477	382
500	—	—	590	472

FULL-LOAD CURRENTS — 3ϕ, AC INDUCTION MOTORS

* in HP
† in A

Siemens

Circuit Power in Resistive Loads

Ohm's law and the power formula can be applied to circuits in which electrical resistance (such as heating elements or incandescent lamps) is the only significant opposition to the flow of current. Theoretical power can be calculated by using the voltage rating of the load and a measured resistance. See Calculating Power.

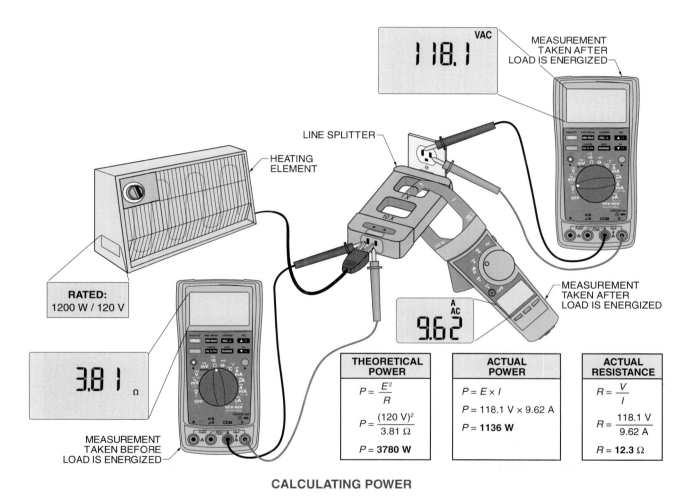

VAC
118.1
MEASUREMENT TAKEN AFTER LOAD IS ENERGIZED

LINE SPLITTER

HEATING ELEMENT

RATED:
1200 W / 120 V

A AC
9.62

MEASUREMENT TAKEN AFTER LOAD IS ENERGIZED

3.81 Ω

MEASUREMENT TAKEN BEFORE LOAD IS ENERGIZED

THEORETICAL POWER	ACTUAL POWER	ACTUAL RESISTANCE
$P = \dfrac{E^2}{R}$	$P = E \times I$	$R = \dfrac{V}{I}$
$P = \dfrac{(120\ V)^2}{3.81\ \Omega}$	$P = 118.1\ V \times 9.62\ A$	$R = \dfrac{118.1\ V}{9.62\ A}$
$P = \textbf{3780 W}$	$P = \textbf{1136 W}$	$R = \textbf{12.3}\ \Omega$

CALCULATING POWER

Actual power output is easily calculated by multiplying voltage by current, which are both measured when the load is energized. Also, resistance calculated from these values is more accurate than directly measured resistance. This is because some loads change resistance significantly when power is applied. For example, a lamp has a much higher resistance when energized than when de-energized. This higher resistance reduces the lamp's actual power output from the lamp's theoretical power output. However, resistance measurements cannot be taken when a circuit is powered, so actual resistance in energized loads must be derived from other measurements.

1. _____ What is the theoretical power output of the load?

2. _____ What is the actual power output of the load?

3. _____ What is the actual resistance of the load?

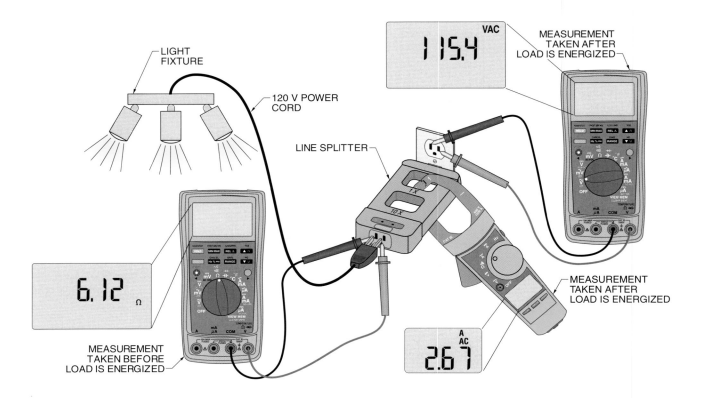

4. _____ What is the theoretical power output of the load?

5. _____ What is the actual power output of the load?

6. _____ What is the actual resistance of the load?

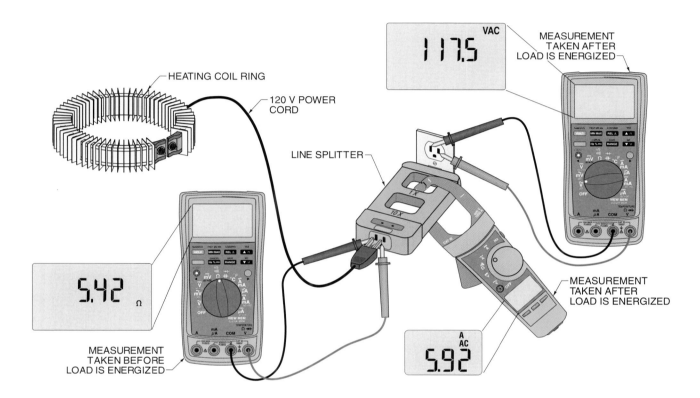

HEATING COIL RING

120 V POWER CORD

VAC

117.5

MEASUREMENT TAKEN AFTER LOAD IS ENERGIZED

LINE SPLITTER

5.42 Ω

MEASUREMENT TAKEN BEFORE LOAD IS ENERGIZED

MEASUREMENT TAKEN AFTER LOAD IS ENERGIZED

A AC

5.92

7. _____ What is the theoretical power output of the load?

8. _____ What is the actual power output of the load?

9. _____ What is the actual resistance of the load?

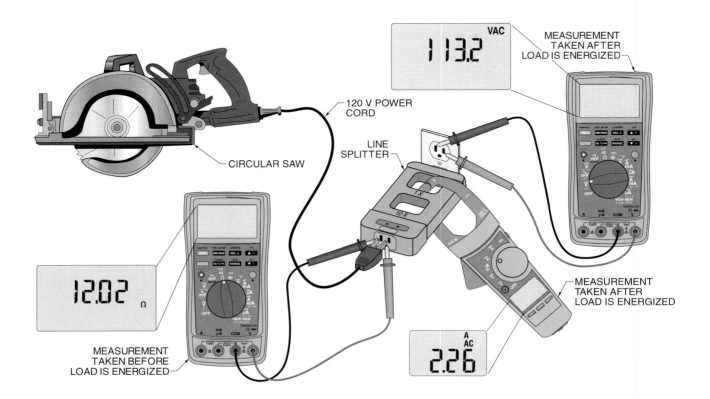

_____ 1. What unit is used to measure frequency?

T   F   2. Higher carrier frequencies produce fundamental frequencies closer to a pure sine wave.

_____ 3. What unit is used to measure the intensity of sound?

T   F   4. Rectifier circuits convert DC to AC.

_____ 5. Which electrical quantity is typically measured when testing mechanical switches?

T   F   6. Larger amplitude sound waves produce louder sounds.

_____ 7. Which type of digital logic gate inverts the incoming signal?

_____ 8. Which electrical quantity is typically measured when testing transistors?

_____ 9. Which type of digital logic gate de-energizes the output if one or more inputs are energized?

T   F   10. Gain is the ratio of output signal to input signal.

_____ 11. Does increasing frequency increase or decrease wavelength?

T   F   12. A manual switch is a switch with no moving contacts.

_____ 13. Does increasing oscilloscope sample rate increase or decrease the accuracy of the waveform trace?

_____ 14. If an oscilloscope measures 17 V with a 10x probe, what is the actual circuit voltage?

T   F   15. The decibel scale is linear.

T   F   16. Sound waves require a medium, such as air, to travel through.

_____ 17. What input condition is prevented by a pull-up resistor?

_____ 18. Which type of solid-state switch does not require the control current to remain ON for the switch to remain ON?

_____ 19. Which type of digital logic gate energizes the output only if all inputs are energized?

_____ 20. Digital logic probes measure a high or low state of what electrical quantity?

21. How can dual-trace oscilloscopes monitor signal gain?

22. What is the difference between fundamental frequency and carrier frequency?

23. Which three types of waves can be output by signal generators?

24. What are the two primary applications of transistors?

25. Explain at least one advantage and one disadvantage of high carrier frequencies in motor drives.

Electronic Circuit Test Instruments 5

Name_____ Date_____

Line Splitters

Taking current measurements requires either opening the circuit for an in-line current measurement (usually 10 A or less), or using a clamp-on current ammeter or attachment. It is usually easier to take clamp-on meter measurements than in-line measurements because they do not require opening or modifying the circuit in any way. Another advantage of using a clamp-on ammeter or attachment is that they can measure several hundreds (or thousands) of amps without the potential safety hazard of opening the circuit.

There are two disadvantages to using a clamp-on ammeter for current measurements. First, clamp-on ammeters are generally not designed to take small current measurements (typically less than 1 A). The second is that clamp-on ammeters must be enclosed around only one conductor. This can be difficult when trying to measure the current of loads such as computers, power tools, portable heaters, or any other load that uses a two- or three-conductor power cord.

A line splitter is a measurement accessory that is designed to make measuring current easier for loads that are low-current devices or for when it is difficult to find a single-conductor measuring point. Most line splitters include a 1x slot and a 10x slot where the current clamp can easily be placed. When placed in the 1x slot, the ammeter displays the actual circuit current and is used for measurements 1 A or higher. When placed in the 10x slot, the ammeter displays 10 times the actual circuit current and is used for measurements less than 1 A. By multiplying the current by 10, the clamp-on meter can take a more accurate measurement. When using the 10x slot, the meter reading must be divided by 10 in order to get the actual current measurement. See Line Splitter.

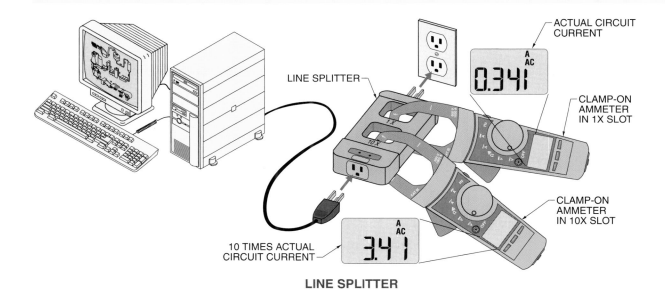

LINE SPLITTER

129

When using a clamp-on attachment accessory on a multimeter, careful attention must be paid to the output type. Some clamp-on attachments output 1 mA/A and others output 1 mV/A. For example, an actual current of 2.45 A will read as 2.45 mA on some clamp-on attachments and 2.45 mV on others. The clamp's output type, mA or mV, will also determine the correct setting and connection of the multimeter.

1. Connect the clamp attachment to the proper meter jacks on the multimeter.

2. Set (draw in) the function switch on the multimeter to the correct position.

3. _____ If the meter reads 84.35, what is the actual current draw (in A) of the load under test?

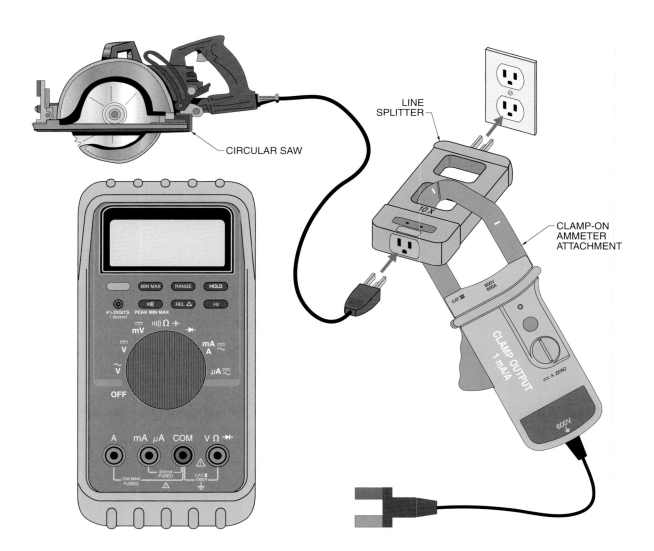

4. Connect the clamp attachment to the proper meter jacks on the multimeter.

5. Set (draw in) the function switch on the multimeter to the correct position.

6. _____ If the meter reads 0.097, what is the actual current draw (in A) of the load under test?

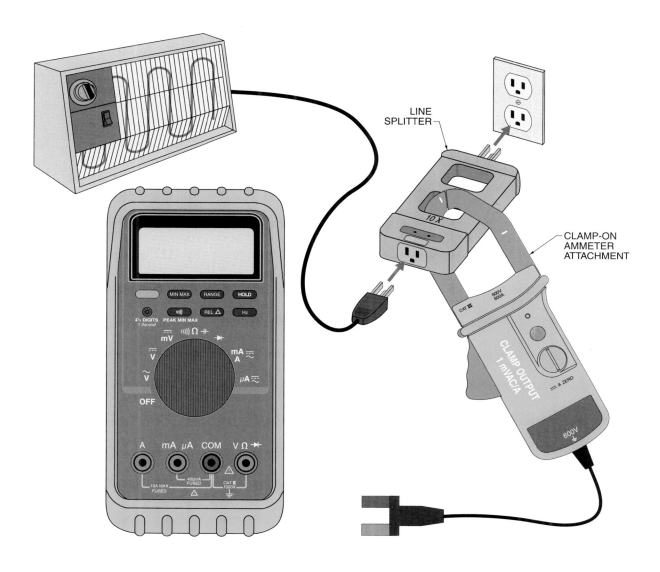

When AC powers DC equipment, diodes are used to convert the AC into DC. The most familiar examples of this are the power converters used to recharge cell phones or laptop computers from 120 VAC receptacles. The power converters are plugged into a 120 VAC receptacle and deliver 9 VDC, 12 VDC, or other DC voltages. These are common, and a troubleshooter usually knows when to set a meter for an AC voltage measurement or a DC voltage measurement. A troubleshooting problem exists when the troubleshooter assumes a circuit is AC that is actually DC, or the troubleshooter assumes a circuit is DC that is actually AC. In such cases, the incorrect setting of the meter (set on AC when it should be on DC) will cause measurement errors and incorrect diagnoses about the actual problem.

Several rules can help eliminate measurement error when troubleshooting an unknown voltage type, or when measurements do not seem right.

Voltage Measurement Rules

- Most meters can be set to measure AC voltage and be connected to a DC voltage without damaging the meter, circuit, or circuit components. However, the measured results will not be correct.

- Most meters can be set to measure DC voltage and be connected to an AC voltage without damaging the meter, circuit, or circuit components. However, the measured results will not be correct.

- If the voltage type (AC or DC) is unknown, take a measurement with the meter set to measure DC voltage. Reverse the meter test leads and take the measurement again. If the voltage is DC at the test point, the two measured values will be the same but one will have a negative (−) reading (for example, 12 VDC and −12 VDC).

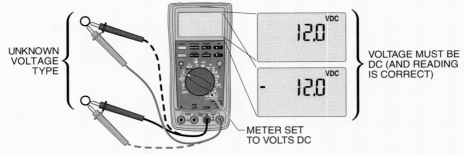

- If the two measured values are both negative and fluctuating (like ghost voltages), the voltage is not DC. Set the meter to measure AC and retake the measurements. If the voltage at the test point is AC, both readings will be the same (for example, 117.5 VAC and 117.5 VAC) and neither one will be negative.

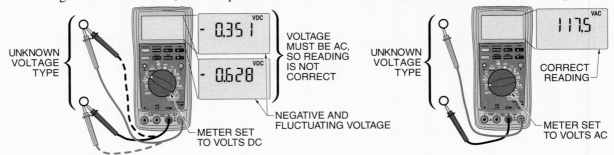

- If the measurements do not clearly indicate whether the voltage at the test point is DC or AC, a graphic display voltmeter (scope type) must be used to observe the voltage waveform. The viewed waveform may show that the voltage includes both AC and DC elements.

Troubleshooting a circuit that includes a solenoid is a good example of test measurements that may not seem correct, even if there is no problem. AC solenoids are rated either intermittent- or continuous-duty. An intermittent solenoid uses AC through a coil to develop a magnetic field and move the solenoid plunger. The AC voltage will heat the coil because it is constantly varying in magnitude and direction. However, the heat buildup will not be a problem if the solenoid is only energized for a short time (usually less than 60 seconds). See Intermittent-Duty Solenoid.

If the solenoid is to be energized for a longer time period, then either the coil must be physically larger to dissipate the heat or the solenoid must use DC instead. Intermittent-duty-rated solenoids are converted to continuous-duty by placing a diode in series with the coil. The diode converts the AC voltage into half-wave DC voltage. Half-wave DC voltage allows for a cooler operating solenoid because the voltage never reverses direction and is only applied during the half-cycle in which the diode allows current to flow through the coil. Since there are still 60 half-cycles of voltage per second (from the 60 Hz AC supply), the coil's magnetic field is still strong enough to hold the solenoid plunger in place. See Continuous-Duty Solenoid.

Since the solenoid's nameplate lists the solenoid voltage as AC (for example 120 VAC) and the diode is usually hidden under the cover of the solenoid, it can be difficult to determine that the AC-rated device is actually operating on DC.

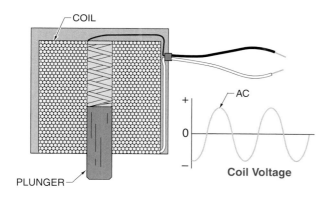

INTERMITTENT-DUTY SOLENOID

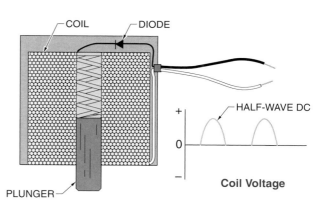

CONTINUOUS-DUTY SOLENOID

An electrician troubleshoots a switch and solenoid in a product handling unit that is operating erratically. Since the solenoid is rated at 120 V/30 W, the multimeters used to measure voltage in the circuit are set to measure VAC.

1. _____ What voltage should Multimeter 2 display when the pushbutton is open?

2. _____ What voltage should Multimeter 2 display when the pushbutton is closed?

3. _____ What voltage should Multimeter 1 display when the pushbutton is closed?

4. _____ Is the voltage between Test Points 2 and 3 AC or DC?

When the pushbutton is open, Multimeter 1 displays unexpected readings that fluctuate between very small voltage values. The electrician suspects that there may be a diode in the solenoid.

5. _____ Is the voltage between Test Points 1 and 2 AC or DC?

6. _____ Which measurement function should be set on Multimeter 1 to measure the correct type of voltage?

7. _____ Will Multimeter 1 read a constant voltage while the pushbutton is open?

8. _____ If the diode shorted out, which type of voltage (AC or DC) would be applied to the solenoid when the pushbutton closed?

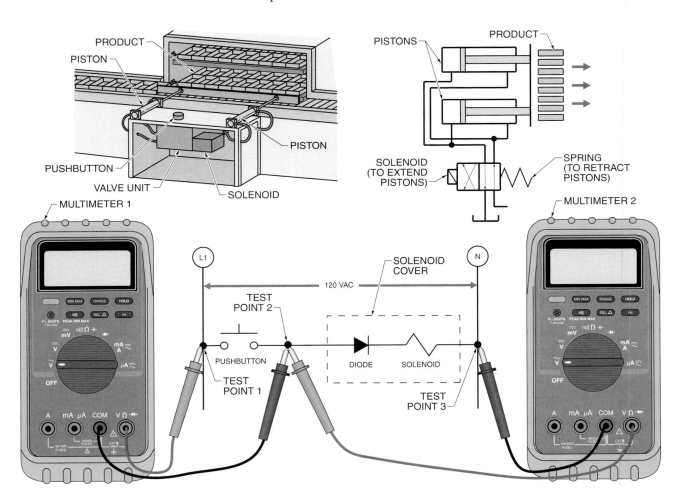

9. Draw the waveform pattern that would be viewed on Oscilloscope 1 when the pushbutton is open.

10. Draw the waveform pattern that would be viewed on Oscilloscope 2 when the pushbutton is open.

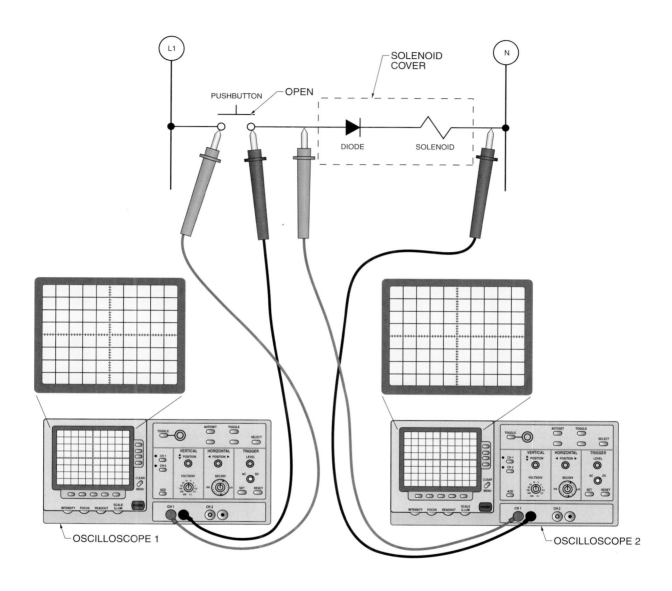

11. Draw the waveform pattern that would be viewed on Oscilloscope 3 when the pushbutton is closed.

12. Draw the waveform pattern that would be viewed on Oscilloscope 4 when the pushbutton is closed.

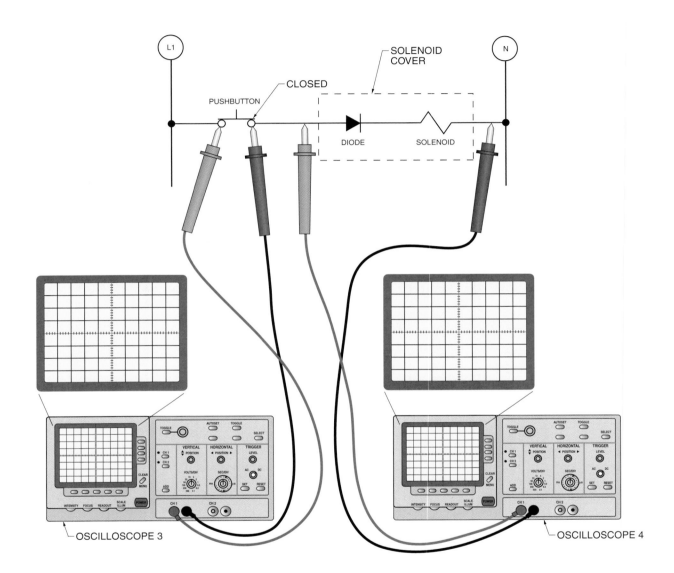

Application 5-3   *Automotive Charging System Testing*

At the heart of the automobile's electrical and electronic systems is a power supply that must deliver the required power (voltage and current) at all times. The battery is the initial power source to start the automobile and supplies enough power to keep some electronic systems operational when the automobile is not running. However, once the automobile is running, it is the alternator that delivers enough power to operate all the electrical devices and recharge the battery. The alternator is a 3φ generator that includes a rectifier (diode) circuit to convert the generated AC into DC. A voltage regulator is used to maintain the proper DC voltage level. See Battery Charging Circuit.

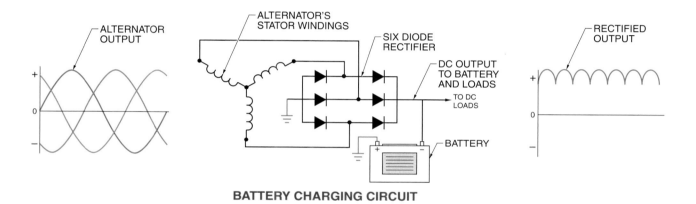

BATTERY CHARGING CIRCUIT

When problems occur in a residential, commercial, or industrial AC electrical system, the power source (power panels, transformers, etc.) is a good place to start troubleshooting. Making sure that the supply power is at the correct voltage and current levels, properly grounded, and not overloaded helps determine potential problems that may be showing up downstream of the power supply. Likewise, a good troubleshooting starting point in automobile electrical circuits is to make sure the power source (battery or alternator) is delivering enough power and at the correct levels at all times. Testing the automobile's charging system will determine the condition of the power supply.

With the engine turned OFF, a meter will display the no-load battery voltage, which should range from 12.4 V to 12.6 V. To test the charging system, the engine is started and brought up to approximately 1500 rpm. A meter set to relative mode will display a positive or negative voltage change, which should range from 0.5 VDC to 2 VDC. If the measured value is higher than 2 VDC, the system is overcharging the battery. If the measurement is a negative value (lower than the no-load voltage) a negative sign appears in front of the measurement and the charging system is not charging the battery.

To fully test the charging system, increase the engine rpm to approximately 2000 rpm and turn ON all major accessories (air conditioning, lights, etc.). The meter display will show the full-load voltage that is above or below the no-load voltage. The full-load voltage should be at least 0.5 V.

1. Set (draw in) the multimeter function switch to measure the voltage at the battery.

2. Connect the multimeter test leads to the proper battery terminal posts.

3. Circle the multimeter function button that zeroes out the no-load voltage measurement.

4. _____ If the no-load voltage measurement was 12.42 V and the meter displays 0.72 V during the 1500 rpm test, what is the actual output of the charging system?

5. _____ If the meter displays −0.43 during the 1500 rpm test, what is the actual output of the charging system?

6. _____ If the voltage measurement decreases to −1.32 V during the 2000 rpm test, what is the actual output of the charging system?

7. What do the negative values indicate?

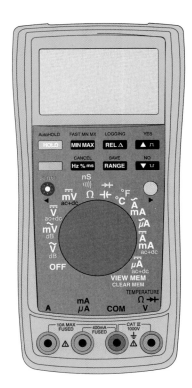

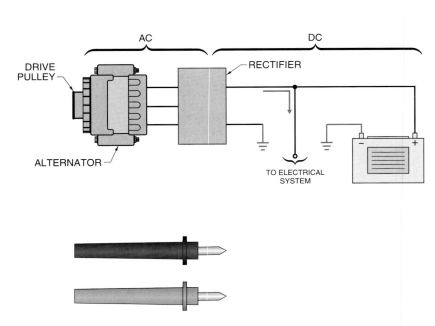

Digital Logic Probes

Digital logic gates (AND, OR, NOT, etc.) are combined to develop various logic functions that can solve almost any control application. Digital logic probes are used to test and troubleshoot digital circuits. In most cases, a digital circuit is replaced (not serviced) if the digital circuit is identified as defective. However, it is important to understand a digital logic system and know how to locate a fault instead of relying on changing boards to solve a problem.

A coin changer circuit is an example of a digital logic circuit. AND and OR gates can be combined to produce the logic required in a basic coin changer. One set of inputs is activated as coins are inserted into a vending machine. Another set of inputs is activated when the cost of the product is determined. If the cost of the product selected is less than the coin inserted, the digital circuit makes the decisions required to release the correct change. For example, if a quarter is inserted and a 10¢ product is selected, the coin changer should give 15¢ in change. To give 15¢ in change, the coin changer must energize CR1 (1 nickel) and CR2 (1 dime). See Coin Changer Circuit.

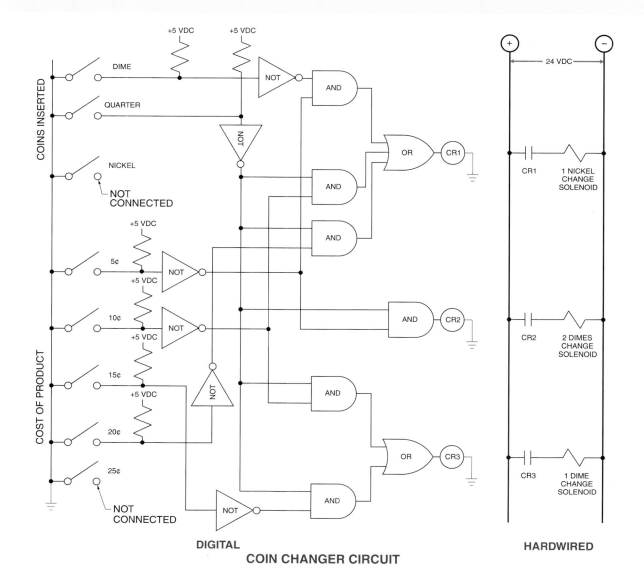

COIN CHANGER CIRCUIT

To simplify the example of a coin changer circuit, it is assumed that only one coin is inserted for the purchase of the product, and the cost of the product selected is 5¢, 10¢, 15¢, 20¢, or 25¢. The NICKEL coin insert switch is intentionally not connected in this simplified circuit because there would not be any change given if only a nickel is used. Likewise, the 25¢ coin product selection switch is also intentionally not connected because there would not be any change given if a 25¢ product is selected and the highest value coin that can be inserted is a quarter.

In a digital circuit, no logic gate can be left floating (not connected to a high or low at all times). To prevent a gate from floating, pull-up resistors are used. The pull-up resistor connects the gate to a HIGH (+5 VDC) as long as the switch is open. When the switch is closed, the gate is connected to ground and thus goes LOW (0 VDC). The pull-up resistor also limits the amount of current flowing through the circuit when the switch is connected to ground. A NOT gate (also called inverter) is used to change a HIGH to a LOW anytime the switch is open. When the switch is closed and there is a LOW at the inverter input, the inverter delivers a HIGH to the coin changer circuit. The inverter receives 5 VDC from the power supply connected to the digital chip (but not shown on the digital print).

Voltage readings are taken before any coins are inserted into the Coin Changer.

1. _____ What voltage should Multimeter 1 display?

2. _____ What voltage should Multimeter 2 display?

A dime is inserted into the Coin Changer and the 5¢ product switch is selected.

3. _____ Should Logic Probe 1 indicate a HIGH or LOW?

4. _____ Should Logic Probe 2 indicate a HIGH or LOW?

5. _____ Should Logic Probe 3 indicate a HIGH or LOW?

6. _____ Should Logic Probe 4 indicate a HIGH or LOW?

7. _____ Should Logic Probe 5 indicate a HIGH or LOW?

8. _____ Should Logic Probe 6 indicate a HIGH or LOW?

9. _____ Should Logic Probe 7 indicate a HIGH or LOW?

10. _____ Should Logic Probe 8 indicate a HIGH or LOW?

A quarter is inserted into the Coin Changer and the 5¢ product switch is selected.

11. _____ Should Logic Probe 1 indicate a HIGH or LOW?

12. _____ Should Logic Probe 2 indicate a HIGH or LOW?

13. _____ Should Logic Probe 3 indicate a HIGH or LOW?

14. _____ Should Logic Probe 4 indicate a HIGH or LOW?

15. _____ Should Logic Probe 5 indicate a HIGH or LOW?

16. _____ Should Logic Probe 6 indicate a HIGH or LOW?

17. _____ Should Logic Probe 7 indicate a HIGH or LOW?

18. _____ Should Logic Probe 8 indicate a HIGH or LOW?

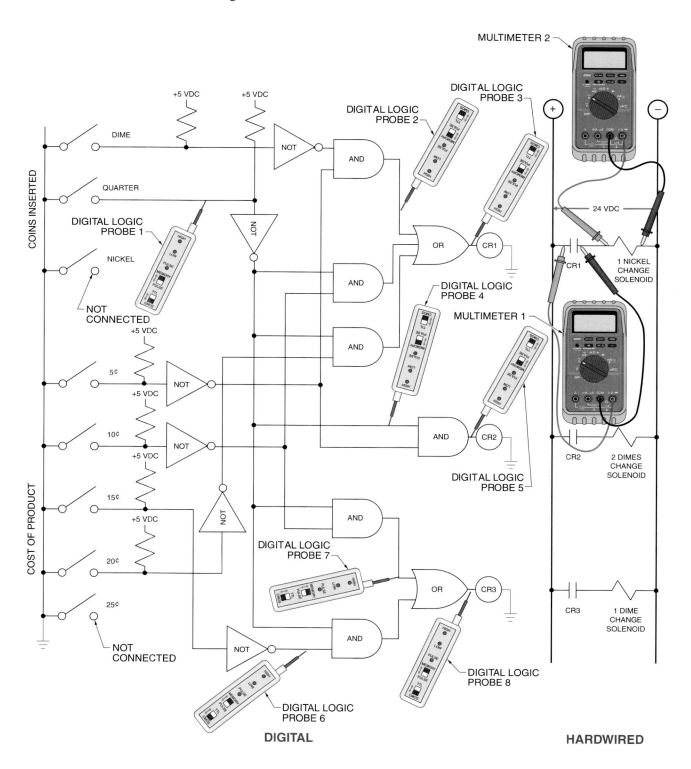

DIGITAL

HARDWIRED

A battery charger recharges DC devices by producing a high current flow through a weak battery and a trickle current flow through a fully charged battery. The battery charger automatically switches from high current flow to trickle current flow when the battery is fully charged. See Battery Charger and Battery Charger Circuit.

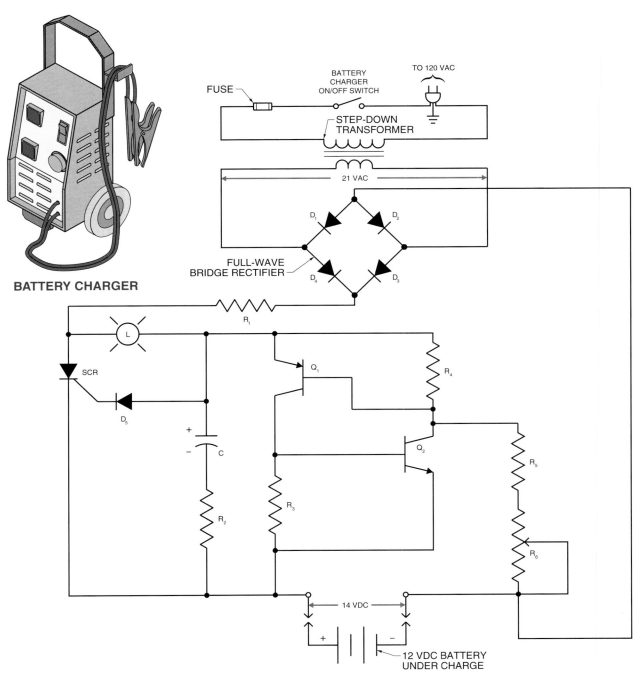

BATTERY CHARGER

BATTERY CHARGER CIRCUIT

In a battery charger designed to recharge 12 VDC batteries, 120 VAC is applied to a step-down transformer when the battery charger ON/OFF switch is closed. A fuse installed in the primary side of the transformer protects the battery charger from excessive current flow. The step-down transformer reduces the supply voltage to approximately 21 VAC. A full-wave bridge rectifier circuit changes (rectifies) the AC into full-wave DC.

A typical battery charger circuit consists of an SCR (silicon controlled rectifier) circuit and a transistor circuit. The SCR circuit controls the high current flow. The transistor circuit controls the trickle current flow. The SCR and transistor circuits are connected in parallel with each other and in series with the battery under charge.

A weak battery connected to the battery charger provides a low resistance current path. The current flows through resistor R_1, the lamp, capacitor C, resistor R_2, and the battery. The capacitor charges and provides enough positive charge on the anode of diode D_5 so that the diode conducts and allows current to flow. The current flowing through the diode triggers the gate of the SCR, which allows high current DC to flow. Current flows through the SCR as long as the weak battery under charge is drawing a high current. The transistor circuit acts like an open switch when the SCR circuit is ON.

A fully charged battery produces a higher voltage in the transistor circuit. This triggers the transistor into conduction, which allows the capacitor to discharge through the transistor circuit. The discharging of the capacitor lowers the voltage at D_5 to the point where it does not allow current to flow to the gate of the SCR. Then the only path for current flow is through the transistor circuit.

The transistor circuit allows only a trickle current to flow because the circuit has a higher resistance than the SCR circuit. The voltage drop across the lamp is great enough to allow the lamp to turn ON when the transistor circuit is conducting. The lamp remains ON as long as the battery under charge is connected into the circuit. The point at which the charging circuit switches to a trickle current flow is set with R_6.

R_1 is placed in series with the charging circuit to limit the current flow if a dead battery is placed on the battery charger. The limited current prevents damage to the battery charger's circuit.

A weak battery is connected to the battery charger and the charger is switched ON.

1. Draw the waveform that should appear on Oscilloscope 1.

2. Draw the waveform that should appear on Oscilloscope 2.

3. Draw the waveform that should appear on Oscilloscope 3.

4. Draw the waveform that should appear on Oscilloscope 4.

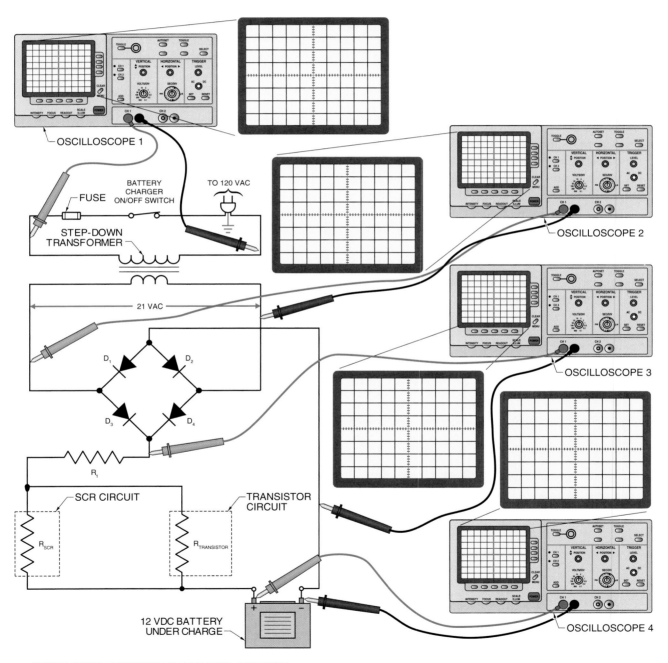

SIMPLIFIED BATTERY CHARGER CIRCUIT

Application 5-6   *Troubleshooting Photoelectric or Proximity Switches*

Solid-state photoelectric and proximity switches are designed to last a long time if properly used. Photoelectric and proximity switches do fail if they experience high transient voltages, draw higher than rated current, or operate in high ambient temperatures. Solid-state photoelectric and proximity switches can be tested using a voltmeter.

When DC photoelectric or proximity switches are required to operate an AC load, a solid-state relay (SSR) is used as the interface between the DC circuit and the AC circuit. The switch is used to operate the SSR input and the SSR output controls the AC load.

Photoelectric and proximity switches are available with either a PNP transistor output (current source, positive switching) or an NPN transistor output (current sink, negative switching). See PNP Transistor Switch and NPN Transistor Switch. The wiring diagram on the switch is used to identify the type of switching used. To determine the type, the placement of the load (L) on the diagram indicates whether the load is connected directly to the positive of the DC power supply at all times (NPN-type switch), or the load is connected directly to the negative of the DC power supply at all times (PNP-type switch). Photoelectric and proximity switches are available with a normally open (NO) output, normally closed (NC) output, or both an NO and NC output in the same switch.

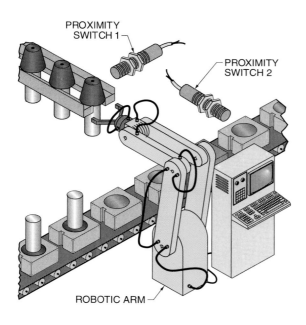

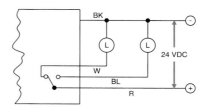

PNP TRANSISTOR SWITCH

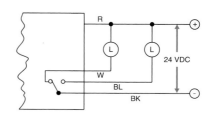

NPN TRANSISTOR SWITCH

Proximity Switch 1 is not activated (no target in front of it).

1. _____ Is the proximity switch an NPN or PNP-type switch?

2. _____ What is the expected voltage reading for Multimeter 1?

3. _____ What is the expected voltage reading for Multimeter 2?

4. _____ What is the expected voltage reading for Multimeter 3?

Proximity Switch 1 is activated (a target in front of it).

5. _____ What is the expected voltage reading for Multimeter 1?

6. _____ What is the expected voltage reading for Multimeter 2?

7. _____ What is the expected voltage reading for Multimeter 3?

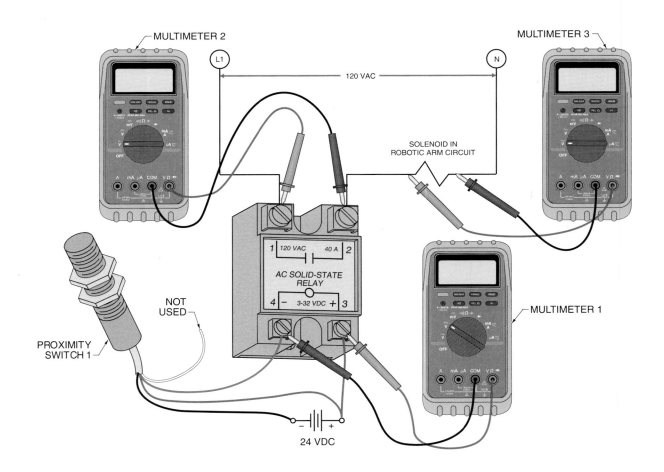

Proximity Switch 2 is not activated (no target in front of it).

8. _____ Is the proximity switch an NPN or PNP-type switch?

9. _____ What is the expected voltage reading for Multimeter 1?

10. _____ What is the expected voltage reading for Multimeter 2?

11. _____ What is the expected voltage reading for Multimeter 3?

Proximity Switch 2 is activated (a target in front of it).

12. _____ What is the expected voltage reading for Multimeter 1?

13. _____ What is the expected voltage reading for Multimeter 2?

14. _____ What is the expected voltage reading for Multimeter 3?

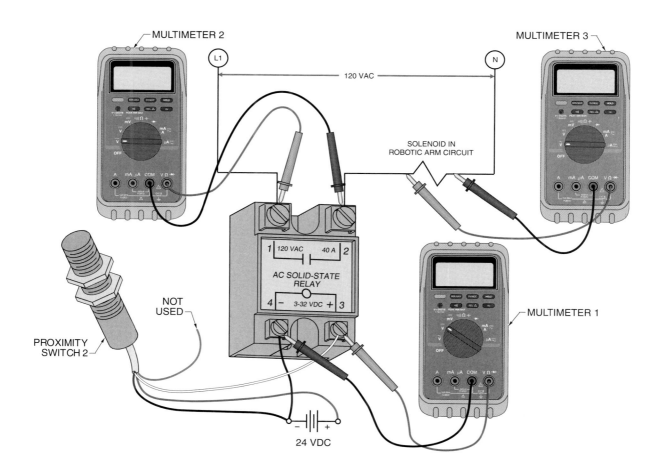

A photovoltaic cell (solar cell) is a device that converts solar energy to electrical energy. Photovoltaic cells produce a voltage when exposed to light. Photovoltaic cells are being used to directly or indirectly power an ever greater number of electronic circuits. Photovoltaic cells can be used to directly power electronic devices such as handheld calculators. However, most photovoltaic cells are used as part of a battery-powered system in which the photovoltaic cells power the electronic circuits when light is present and charge batteries to supply power when light is not available. Applications of photovoltaic cell and battery-powered devices include school crossing or warning signs, portable traffic lights (in construction zones), and remote weather stations.

Photovoltaic cells are rated by the amount of energy they convert. Most manufacturers rate the output in terms of volts (V) and milliamps (mA). Photovoltaic cells follow the same laws as batteries when connected in series and parallel. A circuit of cells connected in series produces a voltage output equal to the sum of the individual cells' voltage outputs, but the current stays the same. For example, 11 cells are connected in series. If each cell produces 100 mA at 0.5 V, then the output of the circuit is 100 mA at 5.5 V. See Photovoltaic Cells in Series.

A circuit of cells connected in parallel produces a current output equal to the sum of the individual cells' current outputs, but the voltage stays the same. For example, if the same 11 cells are connected in parallel, the circuit's output is 1100 mA at 0.5 V. See Photovoltaic Cells in Parallel. By combining series and parallel circuits, any desired voltage and current combination can be designed.

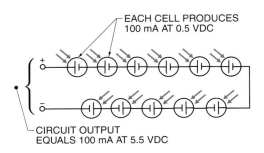

EACH CELL PRODUCES
100 mA AT 0.5 VDC

CIRCUIT OUTPUT
EQUALS 100 mA AT 5.5 VDC

PHOTOVOLTAIC CELLS IN SERIES

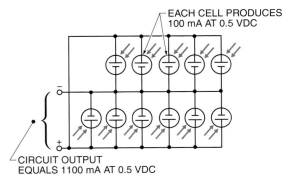

EACH CELL PRODUCES
100 mA AT 0.5 VDC

CIRCUIT OUTPUT
EQUALS 1100 mA AT 0.5 VDC

PHOTOVOLTAIC CELLS IN PARALLEL

Each solar cell is rated for a maximum of 1 VDC at 40 mA.

1. _____ What is the total maximum voltage output (in V) of Circuit 1?

2. _____ What is the total maximum current output (in mA) of Circuit 1?

3. _____ What is the total maximum power output (in W) of Circuit 1?

4. _____ What is the total maximum voltage output (in V) of Circuit 2?

5. _____ What is the total maximum current output (in mA) of Circuit 2?

6. _____ What is the total maximum power output (in W) of Circuit 2?

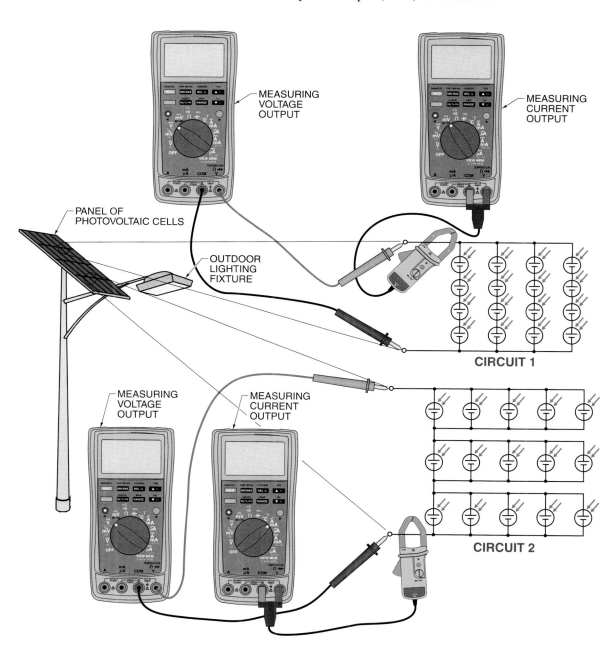

An electric motor drive is an electronic device that controls the direction, speed, torque, and other operating functions of an electric motor, in addition to providing motor protection and monitoring functions. Smaller motor drives (less than 5 HP) are usually not serviced but instead replaced when faulty. Larger motor drives (5 HP and higher) are usually serviced. Individual parts of the drive, such as power supply diodes, capacitors, and large output power transistors can be replaced. Whole sections of a drive, such as the rectifier section, output power transistor section, or control circuit section can also be replaced, eliminating the need to service down to the individual component level.

Before repairing or replacing an electric motor drive, the drive should first be identified as the problem. Test instruments are used to take measurements in the drive system to identify the circuit's problem.

Testing the Power Supply

All electric motor drives have an input voltage rating, which does not need to be the same as the drive's output (and motor) voltage rating. For example, drives less than 5 HP can have voltage input ratings of 120 VAC, 1φ; 240 VAC, 1φ; or 240 VAC, 3φ; and deliver a 240 VAC, 3φ output to drive 3φ motors. The actual input voltage should be within +5% to –10% of the drive's input voltage rating. A voltmeter is used to measure the voltage at the drive's input. See Testing Input Voltage.

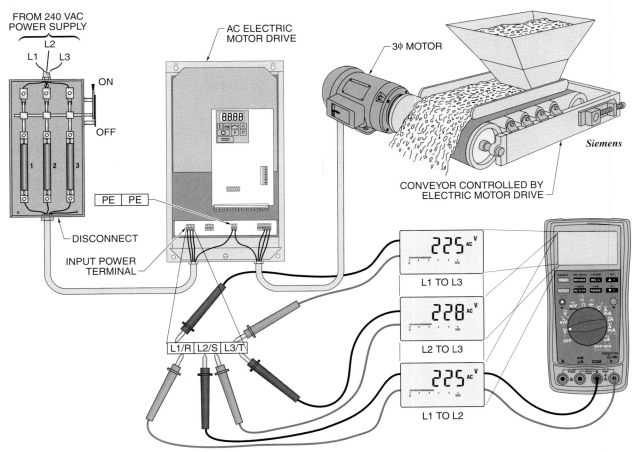

TESTING INPUT VOLTAGE

If the drive does not have the correct voltage, or has no voltage, at its input terminals, test the fuses or circuit breakers protecting the drive. Begin by measuring the voltage coming into the fuses. If the voltage is not correct, troubleshoot upstream from the fuses. If the voltage is within an acceptable limit (+5% to −10%), test the fuses. Fuses are tested one at a time by first measuring the voltage into a fuse and then moving the meter lead from the top of the fuse to the bottom of the same fuse to test the voltage out of the fuse. If voltage is the same out of the fuse as into the fuse, the fuse is good. If there is voltage going into the fuse, but no voltage coming out, the fuse is bad. See Testing Power Supply Voltage.

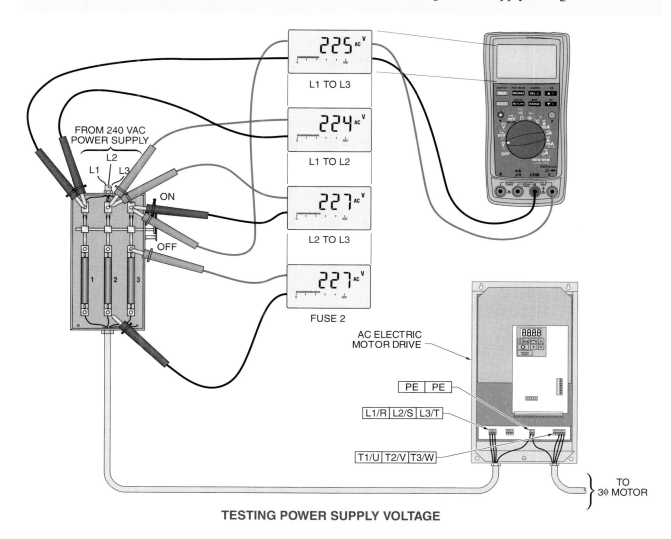

TESTING POWER SUPPLY VOLTAGE

Testing the DC Bus Section

An electric motor drive rectifies the incoming AC supply voltage to DC voltage. The DC bus (link) section of a drive is the rectified (and filtered) DC voltage. The drive inverts the DC back into AC at a controlled (voltage level and frequency) 3ϕ voltage. The DC bus voltage should be 1.4 times the drive's rated output (motor) voltage. Changes in the incoming power supply voltage will proportionally affect the DC bus voltage. For example, if the incoming power supply voltage is 5% less, then the DC bus voltage will be 5% less.

Testing Output Current

It is important to measure the voltage into and out of a drive to make sure the voltage is within an acceptable range. However, measuring current will give a more accurate picture of how much the system is being loaded. If the measured current is equal to the nameplate current rating of the motor, the motor is fully loaded. If the measured current is less than the nameplate current rating of the motor, the motor is underloaded. If the measured current is greater than the nameplate current rating of the motor, the motor is overloaded and/or faulty. The current rating of the drive and the power supply must be equal to (or, preferably, greater than) the highest measured current out of the drive. See Testing Drive Current.

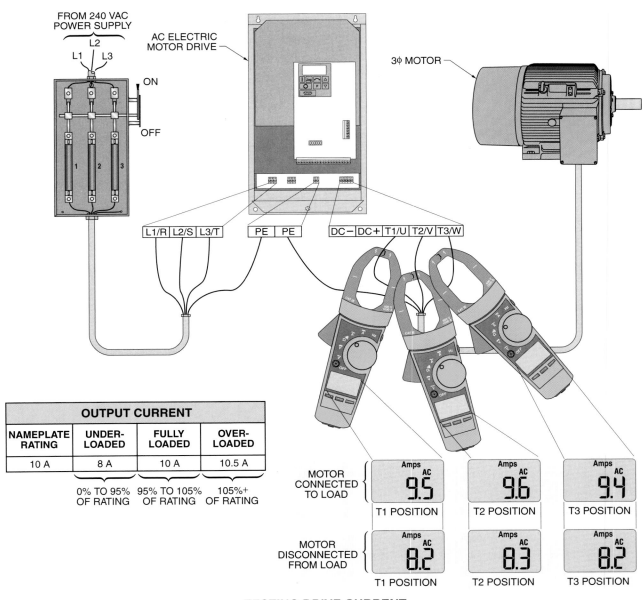

OUTPUT CURRENT			
NAMEPLATE RATING	UNDER-LOADED	FULLY LOADED	OVER-LOADED
10 A	8 A	10 A	10.5 A
	0% TO 95% OF RATING	95% TO 105% OF RATING	105%+ OF RATING

TESTING DRIVE CURRENT

To determine if there is a motor problem or if the motor is overloaded, disconnect the load from the motor and take the current measurements again. The current measurements of an unloaded motor should be less than the motor's nameplate rated current.

The input voltage to a motor drive should be measured when the motor connected to the drive is OFF (no load measurement) and when the motor is fully loaded (full load measurement) to ensure the drive is not overloaded and the power supply is delivering enough power.

1. _____ What is the expected voltage reading from L1 to L3?

2. _____ What is the expected voltage reading from L1 to L2?

3. _____ What is the expected voltage reading from L2 to L3?

4. _____ What is the expected voltage reading from L1 to ground?

5. _____ What is the expected voltage reading from L2 to ground?

6. _____ What is the expected voltage reading from L3 to ground?

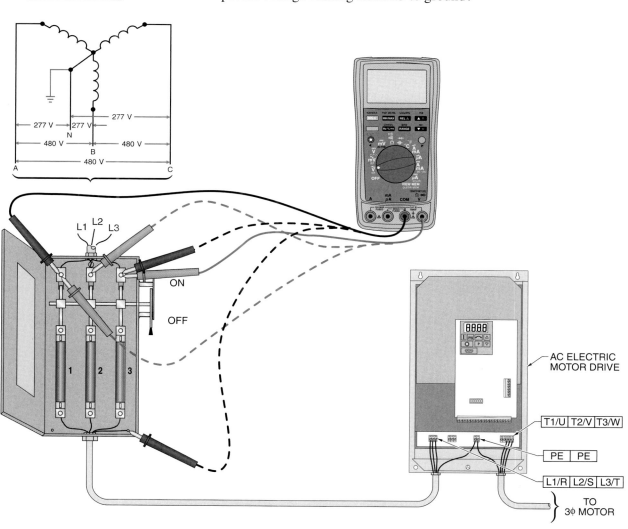

7. _____ What is the percent voltage drop in the drive/motor system?

8. _____ Is this voltage drop in the acceptable range?

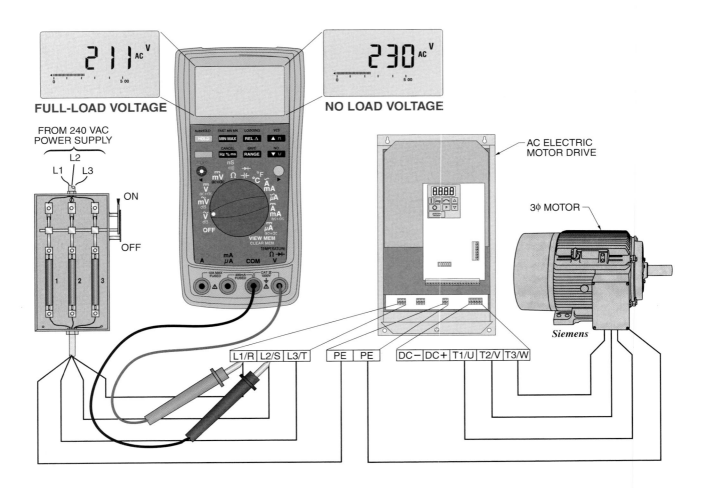

9. _____ If the incoming phase-to-phase voltage is 208 VAC, what is the DC bus voltage?

10. _____ If the incoming phase-to-phase voltage is 220 VAC, what is the DC bus voltage?

11. _____ If the incoming phase-to-phase voltage is 230 VAC, what is the DC bus voltage?

12. _____ If the incoming phase-to-phase voltage is 460 VAC, what is the DC bus voltage?

13. _____ If the incoming phase-to-phase voltage is 480 VAC, what is the DC bus voltage?

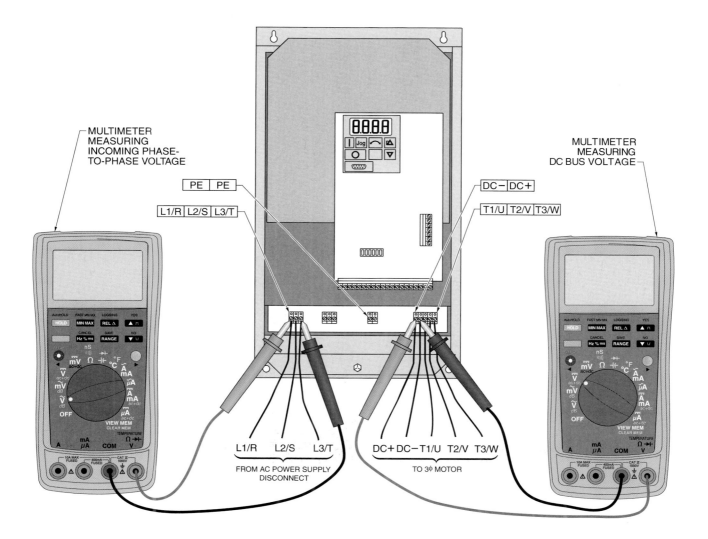

14. _____ What is the expected full-load current reading for Ammeter 1?

15. _____ What is the expected full-load current reading for Ammeter 2?

16. _____ What is the expected full-load current reading for Ammeter 3?

17. _____ What is the expected full-load current reading for Ammeter 4?

18. _____ What is the expected full-load current reading for Ammeter 5?

19. _____ What is the expected full-load current reading for Ammeter 6?

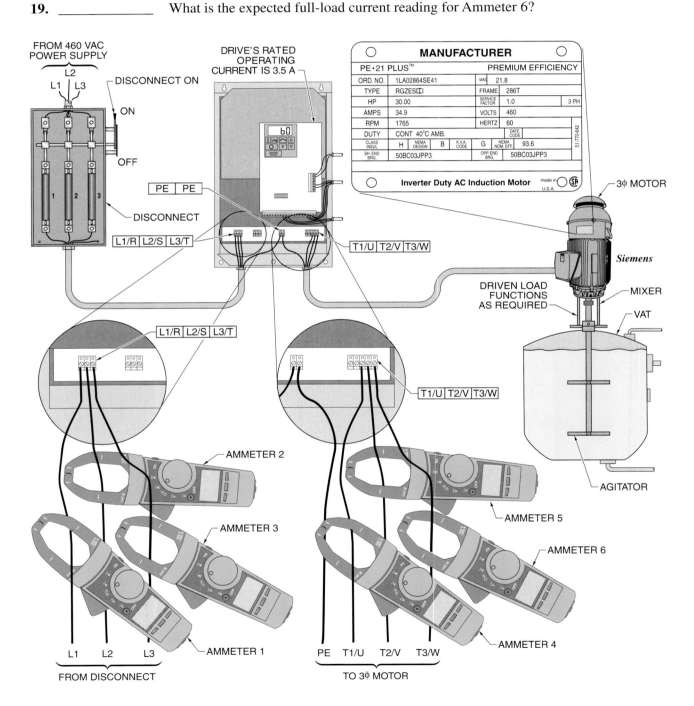

Troubleshooting PLC Inputs and Outputs

A programmable logic controller (PLC) is a solid-state control device that is programmed and reprogrammed to automatically control machines, security systems, lighting systems, and industrial processes. A PLC contains a power supply, input and output modules, processor, and programming terminal. See Programmable Logic Controller.

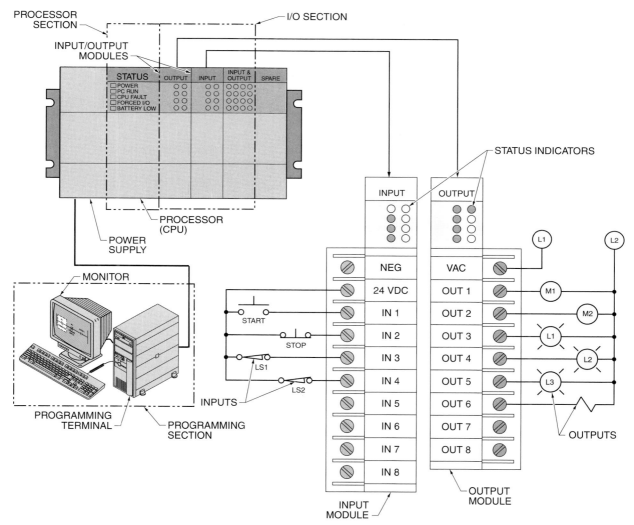

PROGRAMMABLE LOGIC CONTROLLER

The power supply provides necessary voltage levels required for the internal operation of the PLC. The power supply also provides power for the input and output modules. The input and output sections function as the eyes, ears, and hands of the PLC. The input section receives information from pushbuttons, temperature switches, pressure switches, photoelectric and proximity switches, and other sensors. The output section delivers the output voltage required to control alarms, lights, solenoids, starters, and other loads. It is the input and output sections in which most PLC troubleshooting that is not software-related takes place.

The processor section is the brain of the PLC. The processor section is the section of a PLC that organizes all control activity by receiving inputs, performing logical decisions according to the program, and controlling the outputs. The programming section of a PLC is the section that allows input into the PLC program through a keyboard.

1. Set (draw in) the function switch on Multimeter 1 to test the PLC input section.

2. Connect the test leads of Multimeter 1 to the meter and to the PLC input section to test Input 1.

3. _____ What is the expected voltage reading for Multimeter 1 when the switch on Input 1 is open?

4. _____ What is the expected voltage reading for Multimeter 1 when the switch on Input 1 is closed?

5. Set (draw in) the function switch on Multimeter 2 to test the PLC input section.

6. Connect the test leads of Multimeter 2 to the meter and to the PLC input section to test the voltage output of the PLC input module.

7. _____ What is the expected voltage reading for Multimeter 2 if there are no problems?

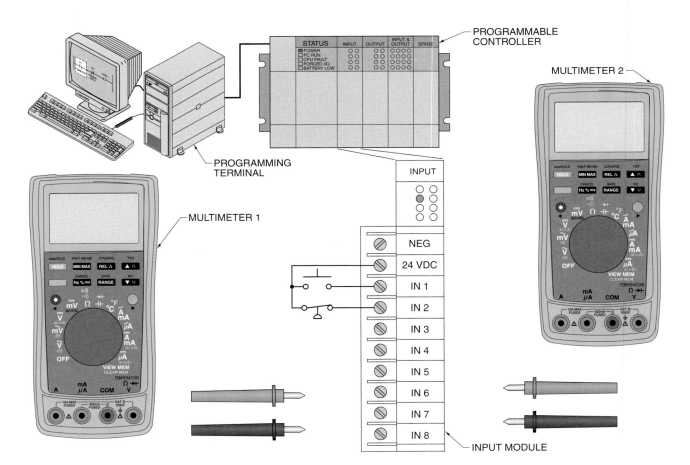

8. Set (draw in) the function switch on Multimeter 3 to test the PLC output section.

9. Connect the test leads of Multimeter 3 to the meter and to the PLC output section to test Output 1.

10. _____ What is the expected voltage reading for Multimeter 3 when the output is energized?

11. Set (draw in) the function switch on Multimeter 4 to test the PLC output section.

12. Connect the test leads of Multimeter 4 to the meter and to the supply voltage for the output loads.

13. _____ What is the expected voltage reading for Multimeter 4 if there are no problems?

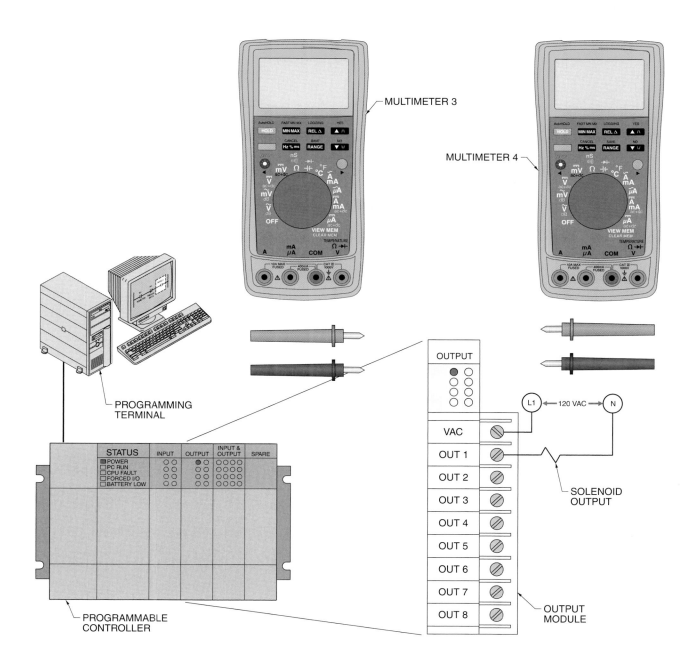

14. Set (draw in) the function switch on Multimeter 5 to test the PLC input section.

15. Connect the test leads of Multimeter 5 to the meter and to the PLC input section to test the proximity switch output.

16. Set (draw in) the function switch on Multimeter 6 to test the PLC input section.

17. Connect the test leads of Multimeter 6 to the meter and to the PLC input section to test the voltage supplied to the proximity switch.

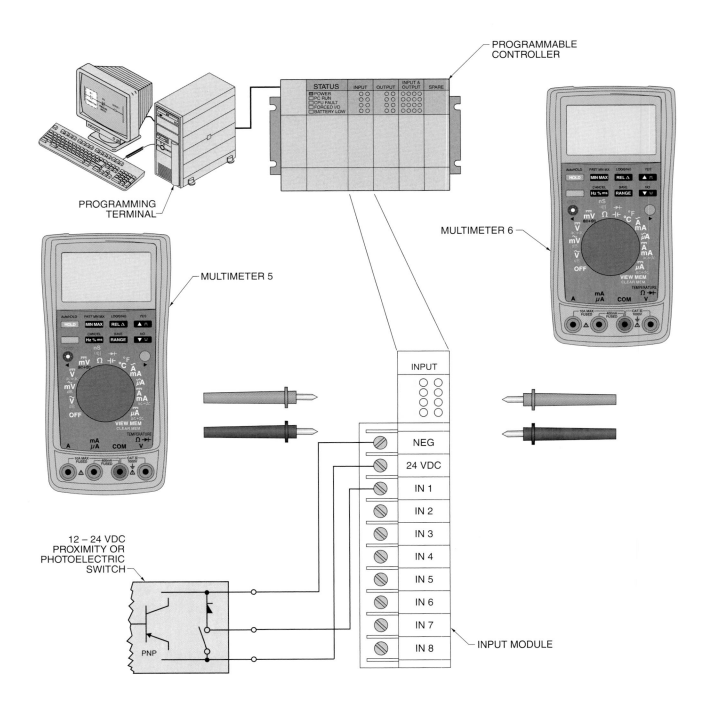

Name _____ Date _____

_____ **1.** What problem is caused by multiple ground connections in a circuit?

_____ **2.** Which type of receptacle disconnects from the power circuit when it detects ground faults?

_____ **3.** What effect does high soil pH have on grounding electrodes?

_____ **4.** Lightning rods are part of which category of grounding?

T   F   **5.** A grounded conductor is usually color-coded green.

_____ **6.** What is the accepted relative distance between Rod 1 and Rod 2 for measuring earth resistance?

_____ **7.** Which type of component is particularly sensitive to electrostatic discharge?

_____ **8.** What is the conductor that connects electrical equipment to the grounding system called?

_____ **9.** What is the maximum grounding electrode resistance required by the NEC®?

T   F   **10.** A grounding system is designed to provide a safe path for fault current to flow to ground.

T   F   **11.** Electronic grounding reduces the noise in electronic signals.

_____ **12.** Which category of grounding includes connecting electrical components to the grounding system?

_____ **13.** What type of soil measurement determines the best material for grounding electrodes?

T   F   **14.** Neutral to ground connections can be made at subpanels.

_____ **15.** What type of clamp-on ammeter is capable of measuring current as low as a few milliamps?

_____ **16.** What unit is used to measure ground resistance?

T   F   **17.** The grounding system for circuits that include sensitive electronic equipment must be significantly less than 25 Ω.

_____ **18.** What is the name of the conductor that connects the grounded parts of the electrical system to the grounding electrode?

_____ **19.** Which two buses are connected by the main bonding jumper?

_____ **20.** What will a voltmeter measure between the hot and ground conductors of a properly grounded 115 VAC receptacle?

21. Why must grounding systems have low resistance?

22. How does an isolated ground receptacle reduce electrical noise?

23. What are examples of existing grounding electrodes?

24. Explain how the rod spacing in an earth resistance test affects the measurements.

25. What are examples of typical causes of fault currents?

Grounding Systems and Earth Ground Test Instruments

6

Name_____ Date_____

System Grounding

Whenever working on or around an electrical circuit or system, the system should be checked to ensure it is grounded. Grounding electrical systems, circuits, and equipment makes them safer by helping to prevent electrical shocks and fires. See Grounding System.

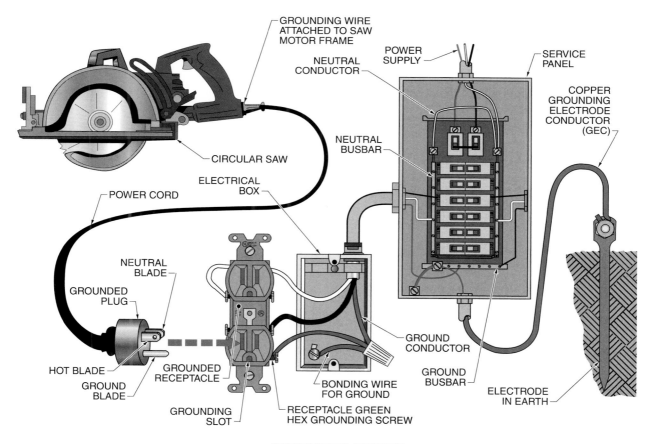

GROUNDING SYSTEM

If an electrical system, circuit, or piece of equipment is grounded, a voltage is present between a hot conductor (ungrounded energized conductor) and ground. A hot conductor is designed to carry current to loads in an electrical circuit and has a voltage potential between it and earth ground equal to the circuit's voltage, such as 120 VAC, 208 VAC, or 480 VAC. Hot conductors are usually colored black, red, blue, orange, brown, or yellow, though black is the most common. Hot conductors should always be properly fused to protect the circuit's conductors and loads from overloads.

A grounded conductor is designed to carry current away from loads in electrical circuits and has been intentionally grounded at a designated location in the electrical system. The grounded conductor is typically called the neutral conductor and is usually colored white or natural gray. Unlike the hot conductor, the grounded (neutral) conductor is not fused.

Grounding connects all exposed non-current carrying metal parts of a system, circuit, or equipment to the earth. Unlike the hot conductor and grounded conductor, the grounding conductor does not carry any circuit or load currents unless there is a problem. The grounding conductor is used to provide a low-resistance path for any fault current to flow to earth ground. Green and green with a yellow stripe are the standard insulation colors for grounding conductors, though grounding conductors can also be bare (no insulation). Grounding conductors, like grounded conductors, are not fused.

In order to test the grounding of a system, circuit, or piece of equipment, an understanding of the different types of services is required. Testing grounding requires knowing the different types of services and how to take voltage measurements with test instruments.

120/240 V, 1φ, 3-Wire Service

A 120/240 V, 1φ, 3-wire service is commonly used to supply power to residential or light commercial buildings. This service provides 120 V, 1φ; 240 V, 1φ; and 120/240 V, 1φ circuits. In residential applications, this service is commonly used for lighting and small appliance use. In commercial applications, this service is commonly used for office equipment, commercial refrigerators, hotel hot tubs and saunas, motors less than 5 HP, cooking equipment, and security equipment. When using many high-power devices, a large power panel or additional power panels may be used. See 120/240 V, 1φ, 3-Wire Service.

120/208 V, 3φ, 4-Wire Service

A 120/208 V, 3φ, 4-wire service is the most common service used for commercial buildings such as offices and schools. It is used to supply customers that require a large amount of 120 V, 1φ power; 208 V, 1φ power; or low-voltage 208 V, 3φ power. This service includes three ungrounded (hot) lines and one grounded (neutral) line. Each hot line has 120 V to ground when connected to the neutral line.

The 120 V circuits are balanced to equally distribute the power from the three hot lines by alternately connecting the 120 V circuits to the power panel so that the phases (A to N, B to N, C to N) are divided among individual load circuits. Likewise, 208 V, 1φ loads, such as 208 V lamps and heating appliances, should also be balanced between phases (A to B, B to C, C to A). Three-phase loads, such as heating elements designed for 3φ power, can be connected to phases A, B, and C. See 120/208 V, 3φ, 4-Wire Service.

120/240 V, 1φ, 3-WIRE SERVICE

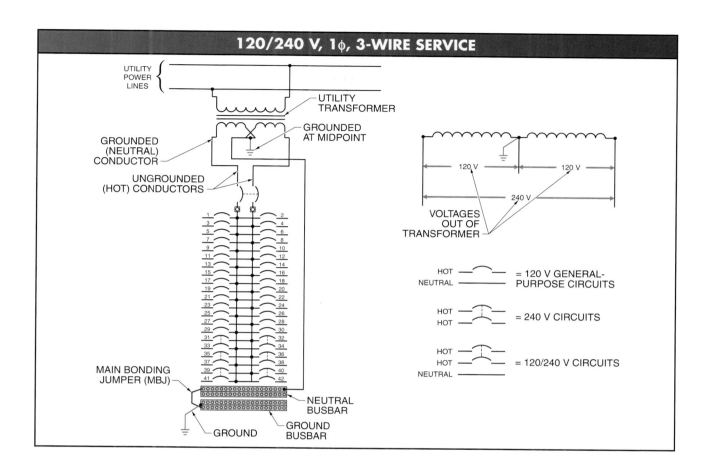

120/208 V, 3φ, 4-WIRE SERVICE

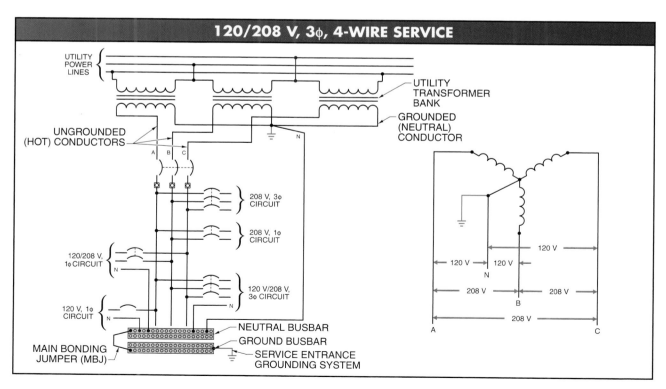

120/240 V, 3φ, 4-Wire Service

A 120/240 V, 3φ, 4-wire service is common in commercial and industrial applications. It is used to supply customers that require a large amount of three-phase power with some 120 V and 240 V, 1φ power. Single-phase power is delivered by one of the three transformers and three-phase power is delivered by using all three transformers. The 120 V, 1φ power is provided by center tapping one of the transformers. Because only one transformer delivers all of the 120 V power, this service is used in applications that require mostly three-phase power or 240 V, 1φ power. In many applications, the total amount of 120 V power used is small when compared to the total amount of three-phase power used. Each transformer may be center tapped if large amounts of 120 V power are required. See 120/240 V, 3φ, 4-Wire Service.

Phase B shall be colored orange (or clearly marked) per the NEC® when the switchboard or panelboard is fed from a 120/240 V, 3φ, 4-wire, delta-connected service. There is approximately 195 V between phase B and the neutral. This is considered an unreliable source of power because 195 V is too high for standard 115 V or 120 V loads and too low for standard 230 V or 240 V loads.

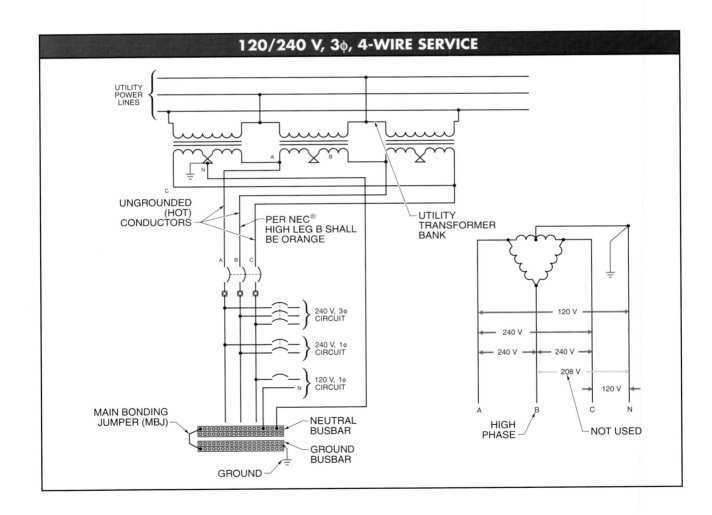

120/240 V, 3φ, 4-WIRE SERVICE

277/480 V, 3φ, 4-Wire Service

A 277/480 V, 3φ, 4-wire service is the same as a 120/208 V, 3φ, 4-wire service, except the voltage levels are higher for industrial applications. This service includes three ungrounded (hot) lines and one grounded (neutral) line. Each hot line has 277 V to ground when connected to the neutral or 480 V when connected between any two hot (A to B, B to C, or C to A) lines. This service provides 277 V, 1φ or 480 V, 1φ power, but not 120 V, 1φ power. Additional transformers can be connected to the 277/480 V, 3φ, 4-wire service to reduce the voltage to 120 V, 1φ. See 277/480 V, 3φ, 4-Wire Service.

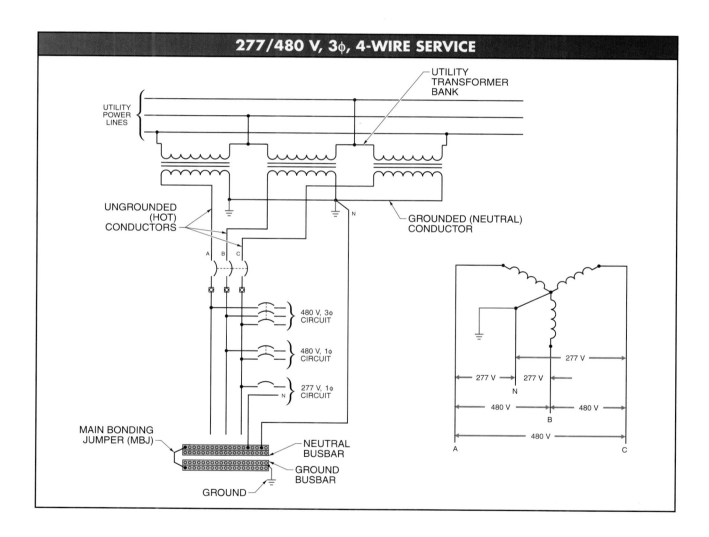

Determine the type of electrical service (voltage levels, phase type, and number of wires) delivered to each service panel and predict the readings of Multimeter 2 and Multimeter 3 if the system is properly grounded.

   1. What type of service is delivered to the service panel?

   2. _____   What should Multimeter 2 read?

   3. _____   What should Multimeter 3 read?

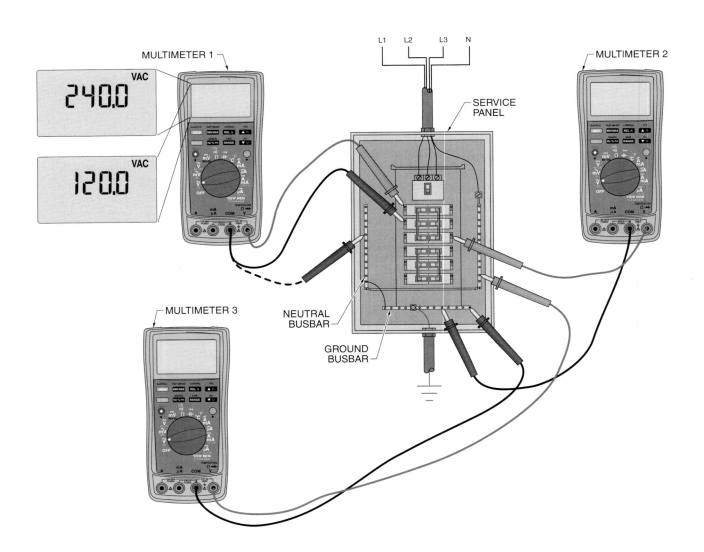

4. What type of service is delivered to the service panel?

5. _____ What should Multimeter 2 read?

6. _____ What should Multimeter 3 read?

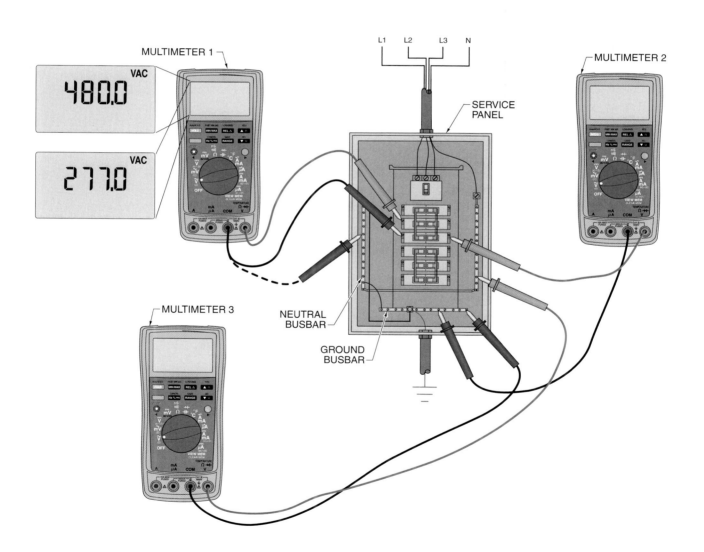

7. What type of service is delivered to the service panel?

8. _____ What should Multimeter 2 read?

9. _____ What should Multimeter 3 read?

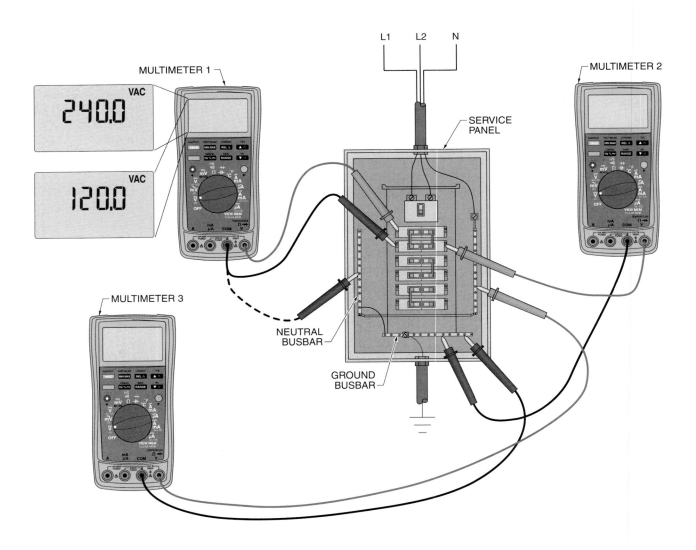

10. What type of service is delivered to the service panel?

11. _____ What should Multimeter 2 read?

12. _____ What should Multimeter 3 read?

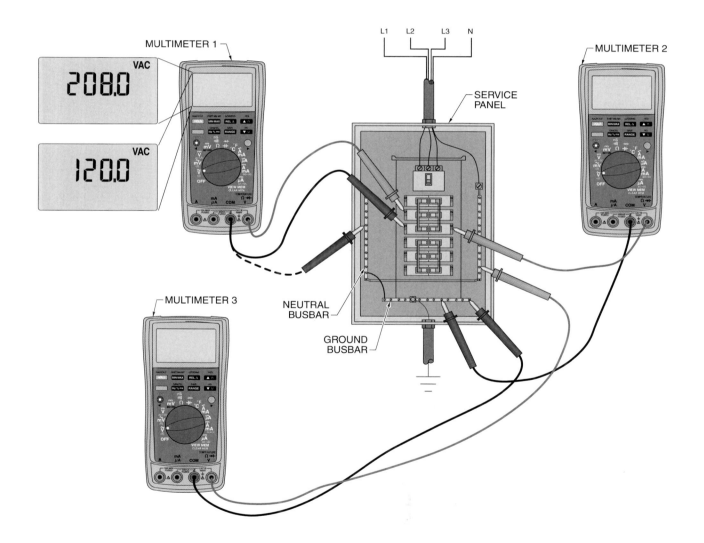

All electrical systems must be grounded by connecting the grounding system to the earth with a grounding electrode, the metal frame of the building, concrete-encased electrodes, a grounding ring, or an underground metal water pipe as per NEC® requirements. Ideally, a solid steel grounding electrode with a minimum diameter of ⅝″ and length of 8′ is driven vertically into noncorrosive soil that has good conductivity (is moist year around). The top of the electrode should be flush with or below ground level, unless protected against physical damage and at least 8′ is in contact with the soil.

In reality, less than optimal soil and ground conditions mean that alternative grounding installations may be needed. Therefore, the NEC® allows some modifications to grounding electrode installation, though the electrode size cannot be reduced and the grounding system resistance must be less than 25 Ω at all times (year-round soil conditions). See Grounding Electrode Installation Requirements.

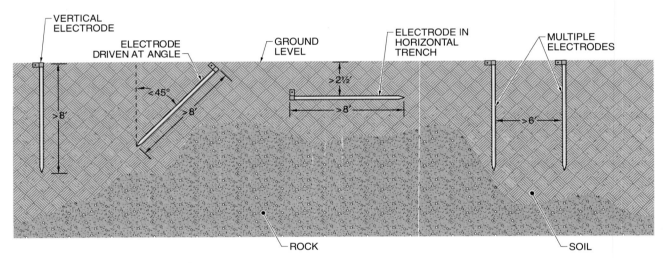

GROUNDING ELECTRODE INSTALLATION REQUIREMENTS

If rocky conditions prevent vertical installation, the grounding electrode can be driven at an angle not to exceed 45° from vertical. Alternatively, the grounding electrode may be buried in a trench that is at least 2½′ deep.

If one electrode exceeds the 25 Ω limit of resistance to ground, additional electrodes can be added to the system to lower the total resistance because of the law of resistance in parallel. However, the rules of calculating resistance in parallel do not apply to multiple grounding rods. The resistance is lowered by set percentages as each additional rod with the same individual resistance is added. The second rod lowers the total resistance to 60% of the first rod. The third rod lowers the total resistance to 40% of the first rod. The fourth rod lowers the total resistance to 33% of the first rod. The multiple electrodes should be at least 6′ apart and connected together at the top.

1. Set (draw in) the function switch position of the ground resistance meter to measure the worst-case leakage current.

2. Set (draw in) the function switch position of the ground resistance meter to measure leakage current of less than 1 A.

3. Set (draw in) the function switch position of the ground resistance meter to measure the grounding system resistance.

4. _____ Is the measured ground resistance acceptable?

5. _____ If a second ground electrode is driven into the ground to lower the resistance, what is the minimum distance that must be maintained between the two grounding electrodes?

6. _____ If the second grounding electrode provides a resistance path equal to the first grounding electrode's resistance, what is the total ground resistance?

7. _____ Is the total resistance of the two grounding electrodes acceptable?

8. _____ If a third grounding electrode provides a resistance path equal to the first grounding electrode's resistance, what is the total ground resistance?

9. _____ Is the total resistance of the three grounding electrodes acceptable?

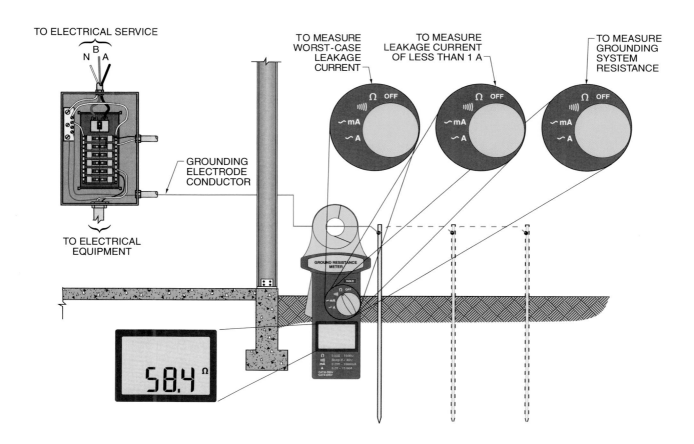

Proper grounding requires that a low-resistance path to earth ground be maintained from all non-current carrying metal throughout an electrical system. The low-resistance ground path can be compromised if the grounding electrode is not properly installed, a ground connection such as a conduit fitting or ground wire splice is no longer providing a low-resistance path to ground, or the ground path has opened.

Losing the ground path by itself does not cause an electrical shock, but does create an electrical hazard in that an electrical shock is now likely to occur. An electrical shock occurs when a person contacts an energized (hot) nongrounded conductor, and either a grounded conductor (neutral) or the ground (non-current carrying metal or ground itself).

A voltmeter can test for proper grounding by connecting between circuit components that should be grounded (all non-current carrying metal and the ground wire) and a known energized (hot) part of the circuit. If the portion of the system under test is properly grounded, the voltmeter will read the system voltage (typically around 115 VAC on standard residential circuits). If there is a partial ground connection (loose or corroded connection), the reading may be any voltage between 0 V and the full system voltage. Lower voltage readings indicate poorer ground connections. See Testing for Ground Faults.

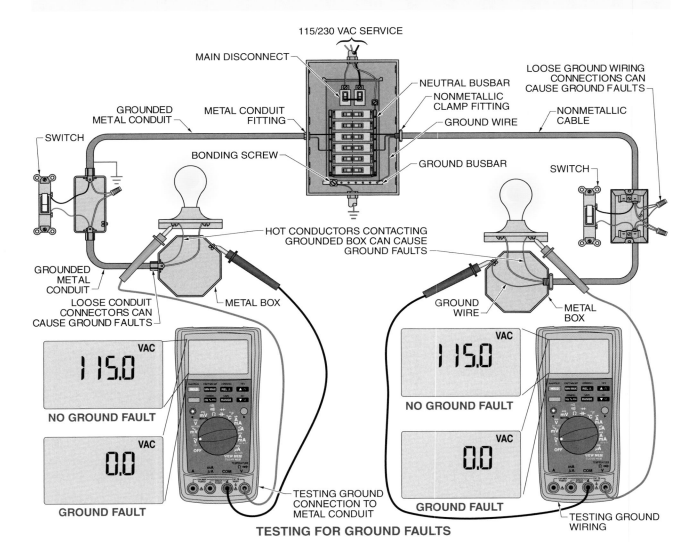

TESTING FOR GROUND FAULTS

1. _____ If the system is properly grounded, what will Multimeter 1 read?

2. _____ If the system is properly grounded, what will Multimeter 2 read?

3. _____ If the system is properly grounded, what will Multimeter 3 read?

A person receives an electrical shock when touching the metal box holding Lamp 1. Testing the system again, Multimeter 1 reads 0 VAC, Multimeter 2 reads 0 VAC, and Multimeter 3 reads 115 VAC.

4. Is the fault most likely located between Lamp 1 and Lamp 2, between Lamp 2 and the switches, or between the switches and the service panel?

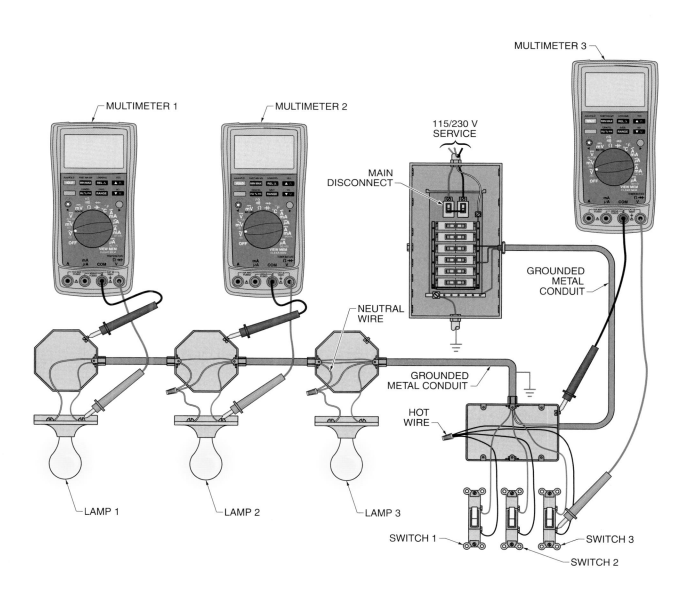

5. _____ If the system is properly grounded, what will Multimeter 1 read?

6. _____ If the system is properly grounded, what will Multimeter 2 read?

7. _____ If the system is properly grounded, what will Multimeter 3 read?

8. _____ If the system is properly grounded, what will Multimeter 4 read?

A person receives an electrical shock when touching the brass switch wall plate of Switch 1. Testing the system again, Multimeter 1 reads 115 VAC, Multimeter 2 reads 115 VAC, Multimeter 3 reads 115 VAC, and Multimeter 4 reads 52 VAC.

9. The most likely location of the fault is between which two components?

10. What would cause the 52 VAC reading on Multimeter 4?

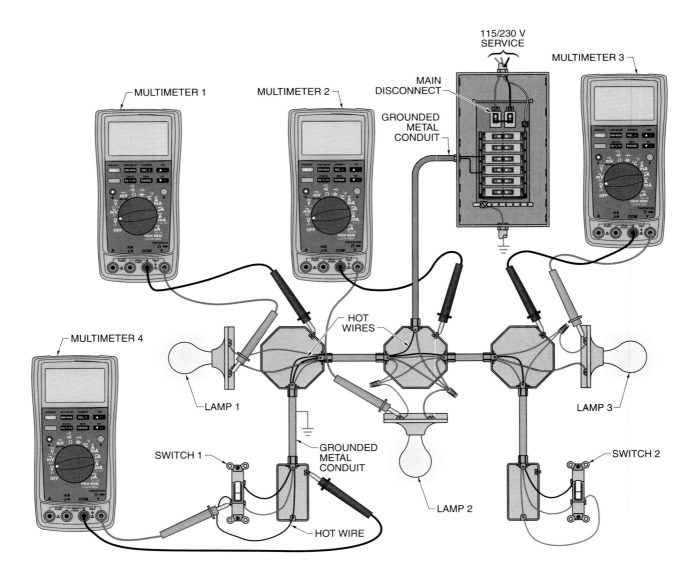

In some circuits, the service entrance grounding conductor may have a small amount of current flowing to ground. Electronic devices such as computers, medical equipment, and sound equipment are highly susceptible to noise introduced into their signals from electromagnetic fields and other sources. The grounding conductor on electronic devices not only prevents electrical shocks but also drains the unwanted noise to ground. This excess current is very small, on the order of a few micro- or nanoamperes, so it probably cannot be measured by a standard clamp-on ammeter. An in-line ammeter set to microamperes or a leakage current ammeter can measure very small ground currents.

Currents flowing from the grounding conductors of multiple electronic devices add together at the service panel and flow to the earth-grounding electrode. Larger electrical systems, especially with more electronic devices, will produce higher total ground currents. For example, a school with many computers will have a higher ground current than a single-family residence. When a measured ground current is higher than expected on a grounding system, additional current measurements should be taken to determine the source of the current. See Measuring Ground Current.

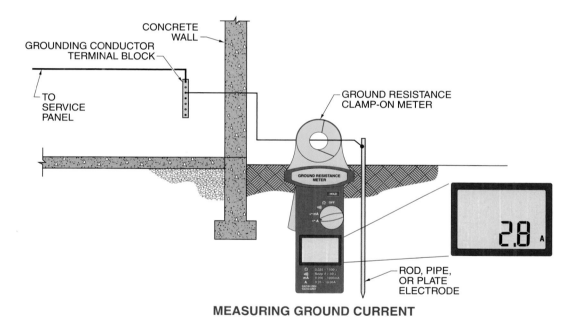

MEASURING GROUND CURRENT

A ground resistance clamp-on meter has measured a higher than expected current (2.8 A) on the outside building grounding system. For each of the given set of clamp-on current measurements taken inside the building, determine the most likely source of the problem from the following list.

A. Load 1
B. Load 2
C. Load 3
D. Other loads connected to the Main Panel
E. Other loads connected to Subpanel 1
F. Other loads connected to Subpanel 2

When taking ground current measurements inside the building, Ammeter 1 reads 3 mA, Ammeter 2 reads 0 mA, Ammeter 3 reads 0 mA, Ammeter 4 reads 2.75 A, and Ammeter 5 reads 1 mA.

1. _____ Where is the most likely source of the ground current?

When taking measurements inside the building, Ammeter 1 reads 3 mA, Ammeter 2 reads 0 mA, Ammeter 3 reads 2.75 A, Ammeter 4 reads 4 mA, and Ammeter 5 reads 1 mA.

2. _____ Where is the most likely source of the ground current?

When taking measurements inside the building, Ammeter 1 reads 3 mA, Ammeter 2 reads 0 mA, Ammeter 3 reads 0 mA, Ammeter 4 reads 4 mA, and Ammeter 5 reads 1 mA.

3. _____ Where is the most likely source of the ground current?

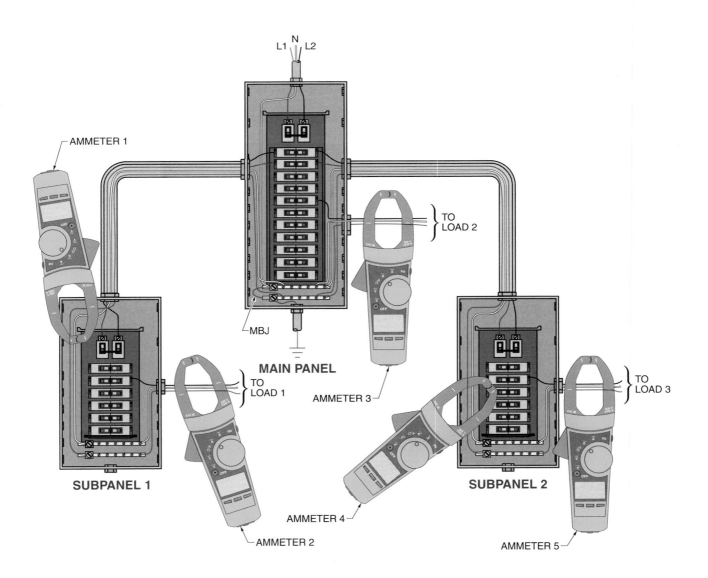

Name_____ Date_____

_____ **1.** Which type of waves do cable height meters use to measure cable heights?

T   F   **2.** A circuit or component must be de-energized before taking resistance measurements.

_____ **3.** Which type of product testing is performed on newly designed components?

_____ **4.** Which type of insulation test is a short-term test that verifies the integrity of insulation?

_____ **5.** Which type of insulation test is similar to a hipot test except that the displayed measurement is in ohms instead of amps?

T   F   **6.** AC hipot tests are performed at 60 Hz.

T   F   **7.** A high voltage test probe accessory displays a voltage on the attached multimeter that is approximately $\frac{1}{1000}$ of the actual voltage.

_____ **8.** Which type of product testing is performed periodically while the product is in service?

_____ **9.** Which type of test instrument is used to verify that a high voltage detector is operating properly?

T   F   **10.** Insulation tests are taken over time (many years) to monitor any slow deterioration.

_____ **11.** Which electrical quantity does a hipot tester measure to determine insulation resistance?

_____ **12.** What is the general rule of thumb for insulation resistance to minimize leakage current?

T   F   **13.** It is normal for insulation resistance to lower gradually over time.

_____ **14.** Which type of product testing is performed after the product is installed but before it is put into service?

_____ **15.** What value calculated from resistance measurements indicates that insulation contains excessive moisture or contamination?

T   F   **16.** A high voltage detector displays the level of voltage present on a power line.

_____ **17.** Which type of hipot test applies voltage in increasing steps?

_____ **18.** Which test instrument takes current measurements of high voltage power lines?

_____ **19.** What will happen at a weak or damaged point in a conductor during a hipot test?

_____ **20.** Which type of test verifies a low impedance path to ground from all exposed metal parts and the grounding conductor?

21. Why is it necessary to test a high voltage detector before and after use?

22. What are several causes of insulation deterioration or damage?

23. Why do some product testing methods require the product to be tested to failure?

24. Why might a standard ohmmeter not accurately measure insulation resistance?

25. How can preventive maintenance such as insulation testing reduce troubleshooting and downtime?

Applications

High Voltage and Insulation Test Instruments 7

Name_____ Date_____

High Voltage Dangers

It is generally understood that working around high voltage is dangerous and that contacting high-voltage lines can be fatal. However, several things about electric shocks are misunderstood. Although it is true that high voltage can kill, it is also true that several thousand volts of static electricity may cause only a harmless shock. It is also possible to receive a fatal electric shock from standard 120 VAC residential circuits and appliances.

Actually, it is the combination of voltage and current that creates the potential for serious electric shock. First, the voltage must be high enough to cause current to flow and, once current flows, the current must be high enough to cause an electric shock.

Voltage causes current to flow once the resistance is low enough. A high resistance between two conducting materials prevents any current flow between them, even when the voltage is high. However, when the resistance decreases, current will start to flow. With high enough voltage, as little as 8 mA of current can cause a painful electric shock. See Effects of Electric Current.

EFFECTS OF ELECTRIC CURRENT	
CURRENT	
20 mA —	Severe muscular contractions, paralysis of breathing, heart convulsions, death
15 mA —	Painful shock; may be frozen or locked to point of electric contact until circuit is de-energized
	Painful shock; removal from contact point by natural reflexes
8 mA —	
1 mA —	Sensation of shock, but probably not painful
0 mA	No sensation

Increasing resistance limits or prevents current flow. Insulating materials such as electrical gloves, rubber-soled shoes, rubber insulating mats, and double-insulated tools all raise the resistance of a potential current path through the body, lowering the possible current flow if a hot (ungrounded) conductor is contacted.

Any voltage above 50 V must be considered dangerous, and all safety rules must be followed regarding personal protective equipment (PPE) and proper tools, equipment, and test instruments. Voltages below 50 V are also considered potentially hazardous when resistance is low enough to cause current to flow. For example, a few volts (6 V, 12 V, 24 V) is usually not enough to cause current flow through a person that has dry hands. This is why a person does not receive an electric shock when touching conductors with small voltages such as small batteries, exposed speaker wires, or computer peripheral cables. However, a person can receive an electric shock if their resistance is lowered enough. For example, people have received electric shocks in showers in which there is a few volts potential between the faucet and drain ground. Thus, measurements should always be taken to determine voltage and current levels.

Determine the amount of current that flows through the body when a person accidentally touches the ungrounded (hot) conductor at a 120 VAC receptacle.

1. _____ How much current (in mA) flows through a body with 50,000 Ω of resistance?

2. _____ How much current (in mA) flows through a body with 40,000 Ω of resistance?

3. _____ How much current (in mA) flows through a body with 30,000 Ω of resistance?

4. _____ How much current (in mA) flows through a body with 20,000 Ω of resistance?

5. _____ How much current (in mA) flows through a body with 15,000 Ω of resistance?

6. _____ How much current (in mA) flows through a body with 10,000 Ω of resistance?

7. _____ How much current (in mA) flows through a body with 5,000 Ω of resistance?

8. _____ How much current (in mA) flows through a body with 3,000 Ω of resistance?

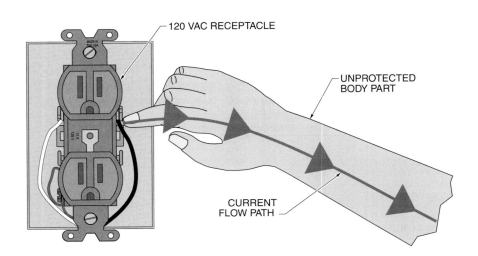

9. Plot the current values against resistance on the graph.

10. _____ At approximately what resistance will this person begin to feel a painful shock?

11. _____ At approximately what resistance may this person be frozen to the point of contact?

12. _____ At approximately what resistance is this person in danger of serious harm or death?

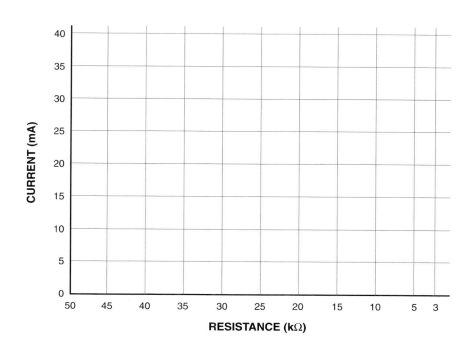

High voltage power lines are run overhead on wood, steel, or reinforced concrete poles; underground in duct lines or conduits; or buried directly in the ground. All high-voltage power lines are dangerous and can electrocute anyone in contact with them. However, accidental contact is more likely with overhead power lines because they are exposed and protected primarily by distance. Higher power line voltages and greater potential for pedestrian or vehicular traffic require that greater minimum clearances be maintained.

The NEC®, OHSA, and other regulating agencies cover minimum clearances (distance) required between electrical conductors and buildings, roofs, driveways, roads, the earth, or anything else on the ground. NEC® 230.24 covers minimum overhead clearance distances of service drop conductors not over 600 V.

Clearance from Ground

The minimum clearance for service drop conductors over the ground (final grade) varies with the voltage of the conductors and the type of traffic underneath. See Clearance from Ground.

- For 150 V or less over pedestrian only traffic (no vehicles), the clearance shall be at least 10′.
- For 300 V or less over pedestrian and car traffic (no trucks), the clearance shall be at least 12′.
- For greater than 300 V but not more than 600 V, over pedestrian and car traffic (no trucks), the clearance shall be at least 15′.
- For not more than 600 V over traffic including trucks, the clearance shall be at least 18′.

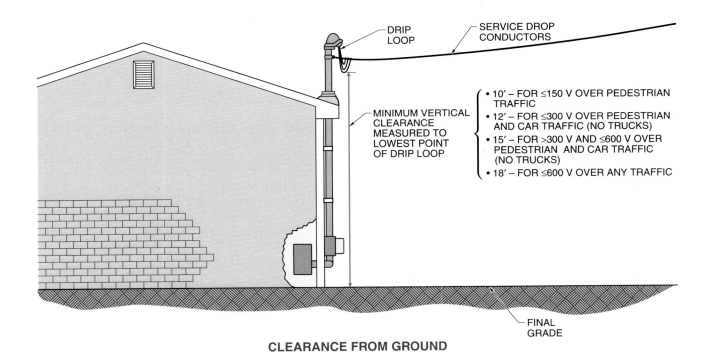

CLEARANCE FROM GROUND

Clearance over Roofs

The minimum clearance for service drop conductors over roofs varies with the voltage of the conductors, the roof type, and conductor termination. See Clearance over Roofs.

- For not more than 600 V over a roof, the clearance shall be 8′. That clearance shall be maintained in all directions for a minimum distance of 3′ from the edge of the roof.
- For not more than 600 V over a roof subject to pedestrian or vehicular traffic, the clearance shall be 18′.
- For not more than 300 V over a rooftop slope of 4″ in 12″ (about 18°) or greater, the clearance shall be 3′.
- For not more than 300 V when the conductors will be terminated at a through-the-roof raceway after not more than 6′ of the conductors pass over the roof, the clearance shall be 18″.

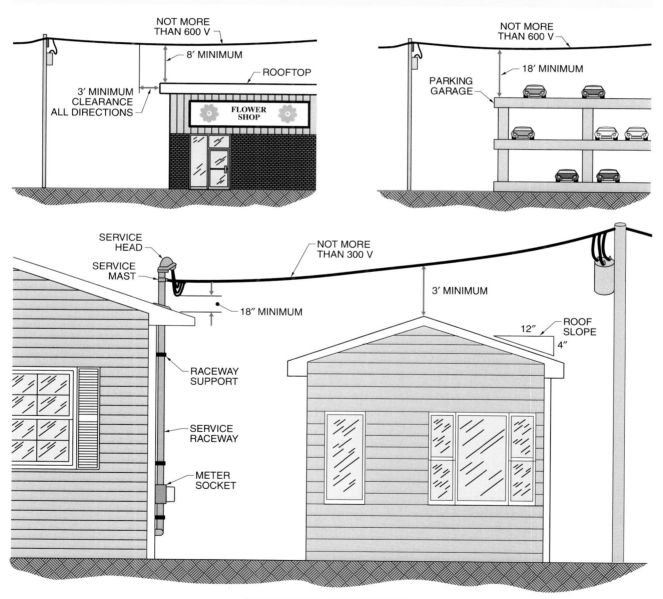

CLEARANCE OVER ROOFS

Even power lines that are properly installed can become dangerous during ice storms, tree fallings, structural additions, and landscape and earth grading changes because the power lines may become closer to the ground or buildings.

Also, power lines can sag because their high current load increases the temperature of the conductors. Sags can be from several feet for shorter spans (150′ to 300′) up to 20′ on long spans (1000′ to 1500′). Power lines usually sag the most on hot days, when electricity demand is at its peak.

For these reasons, the actual distance between a power line and the nearest structure must be measured with a cable height tester if there is any uncertainty about it meeting requirements.

The electric utility installs electrical service to an auto repair shop and measures the height of the installed cable over the surface of the flat roof and over the parking lot. The parking lot traffic consists of pedestrians and small vehicles.

1. _____ Does this installation meet the minimum height clearances for this application?

2. _____ If the shop does a large amount of arc welding, should the cable height be tested again because of the high electrical demand?

3. _____ If the roof is resurfaced by adding 2″ of material, does the installation still meet the minimum height clearance?

4. _____ If the shop begins receiving truck deliveries in their parking lot, does the installation still meet the minimum height clearance?

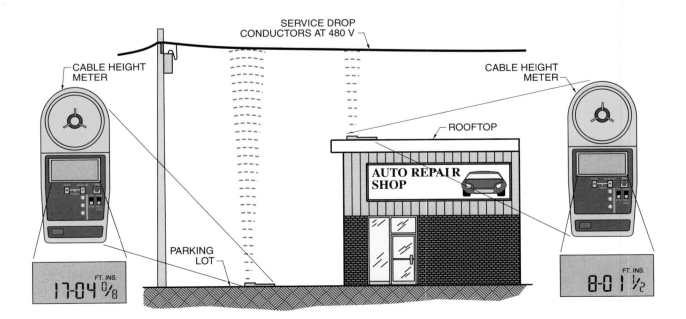

A homeowner wants to build a house addition and a detached garage in the backyard, both underneath the service drop conductors supplying the home.

5. _____ Does the existing installation meet the minimum height clearances for this application?

6. _____ If the garage is a flat-roof design, what is the maximum allowable garage height for that location?

7. _____ If the garage is a sloped-roof design with a 5″ in 12″ slope, what is the maximum allowable garage height for that location?

8. _____ If the addition to the home is on the side with the service mast, will the service mast need to be reconfigured to meet clearance requirements?

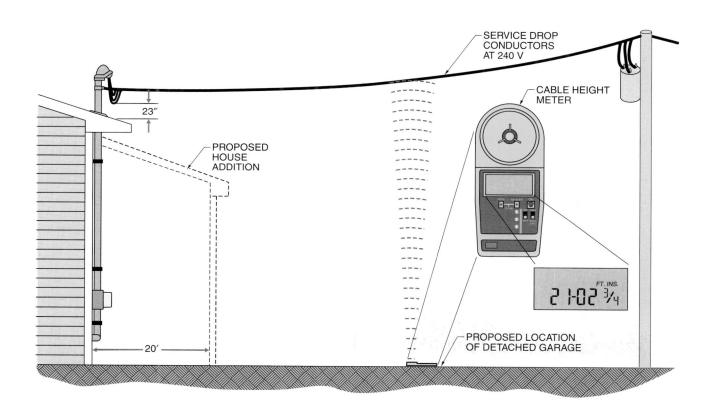

Excessive moisture, chemicals, and dirt cause wire insulation to break down. Insulation breakdown reduces resistance, which allows more leakage current between the windings, or between the windings and the frame of the motor. The excessive leakage current damages the windings and may cause hazardous short circuits.

A megohmmeter can measure the resistance between pairs of motor windings, and between motor windings and the frame of the motor (ground). The resistance measurements are taken with the motor disconnected or the power off and locked out. The amount of leakage current can be calculated by dividing the motor's rated voltage by the resistance measurements. Leakage currents up to 1 mA per 1000 V are considered acceptable.

1. _____ What was the leakage current (in mA) between the winding and the frame when the motor was put into service?

2. _____ What is the leakage current (in mA) between the winding and the frame one year later?

3. _____ Has the insulation deteriorated unacceptably in one year?

4. _____ If the motor was reconfigured for 230 V, what would the leakage current (in mA) be?

5. _____ Would the leakage current in the 230 V configuration be acceptable?

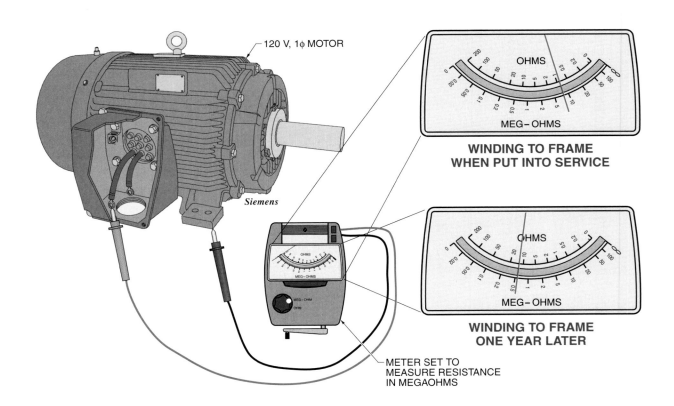

120 V, 1φ MOTOR

Siemens

WINDING TO FRAME
WHEN PUT INTO SERVICE

WINDING TO FRAME
ONE YEAR LATER

METER SET TO
MEASURE RESISTANCE
IN MEGAOHMS

6. _____ What was the leakage current (in mA) between Winding 1 and the frame when the motor was put into service?

7. _____ What was the leakage current (in mA) between Winding 2 and the frame when the motor was put into service?

8. _____ What was the leakage current (in mA) between Winding 3 and the frame when the motor was put into service?

9. _____ What is the leakage current (in mA) between Winding 1 and the frame one year later?

10. _____ What is the leakage current (in mA) between Winding 2 and the frame one year later?

11. _____ What is the leakage current (in mA) between Winding 3 and the frame one year later?

12. _____ Which winding's insulation has deteriorated unacceptably in one year?

13. _____ Which winding's insulation must be monitored for further deterioration?

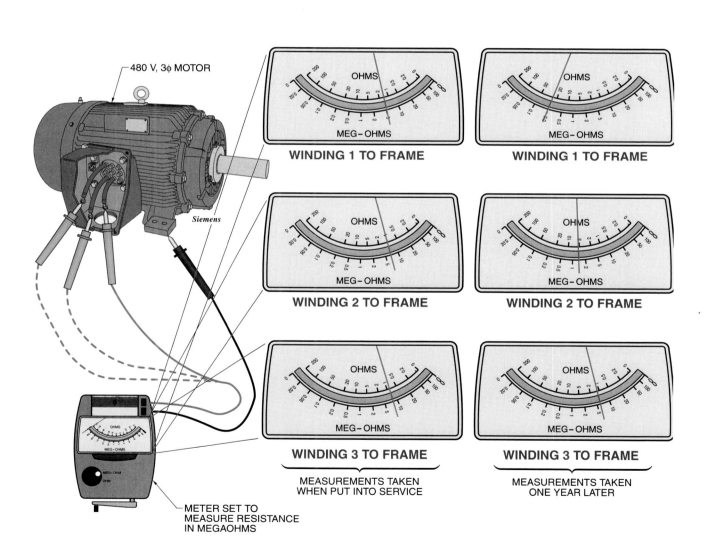

480 V, 3ϕ MOTOR

Siemens

WINDING 1 TO FRAME   WINDING 1 TO FRAME

WINDING 2 TO FRAME   WINDING 2 TO FRAME

WINDING 3 TO FRAME   WINDING 3 TO FRAME

MEASUREMENTS TAKEN WHEN PUT INTO SERVICE   MEASUREMENTS TAKEN ONE YEAR LATER

METER SET TO MEASURE RESISTANCE IN MEGAOHMS

An insulation spot test uses a megohmmeter to measure the resistance of insulation on conductors inside motors. Measurements are taken between the windings and between each winding and ground. The insulation spot tests should be performed when a motor is placed into service and every six months afterward. The measurements are recorded on a chart or plotted on a graph for keeping track of the condition of the motor's insulation over time.

Three common types of DC motors include DC series, DC shunt, and DC compound motors. Each is insulation spot-tested in a similar way, but the internal configuration of the windings must be known in order to test at the proper terminal. A DC series motor has the field windings connected in series with the armature. See DC Series Motor. A DC shunt motor has the field windings connected in shunt (parallel) with the armature. See DC Shunt Motor. A DC compound motor has the field windings connected in both series and shunt with the armature. See DC Compound Motor.

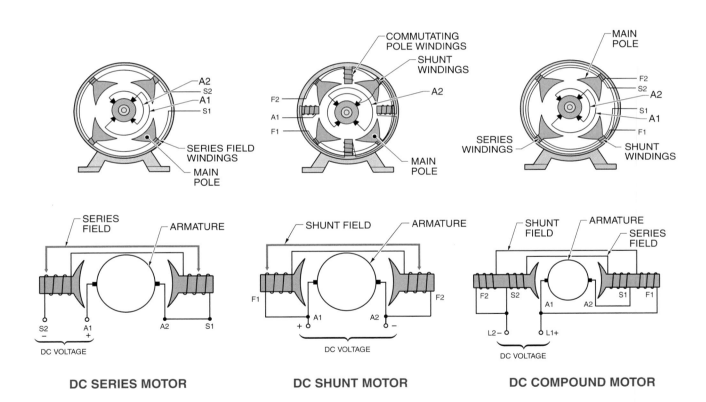

DC SERIES MOTOR   **DC SHUNT MOTOR**   **DC COMPOUND MOTOR**

1. Connect the megohmmeter leads for taking resistance measurements between S1 and ground (the motor frame).

2. Plot the insulation spot test measurements on the insulation spot test graph.

3. What is the problem with the winding insulation?

4. _____ During which time period did the problem begin?

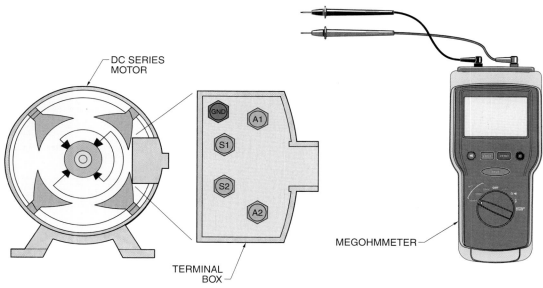

DC SERIES MOTOR

GND A1

S1

S2

A2

TERMINAL BOX

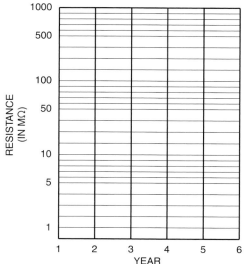

MEGOHMMETER

INSULATION SPOT TEST	
Test Date	**Resistance***
Jan, Year 1	400
Jul, Year 1	400
Jan, Year 2	300
Jul, Year 2	275
Jan, Year 3	250
Jul, Year 3	90
Jan, Year 4	20
Jul, Year 4	1
Jan, Year 5	0.5
Jul, Year 5	—
Jan, Year 6	—

 * in MΩ

RESISTANCE (IN MΩ)

1000
500

100

50

10

5

1

1 2 3 4 5 6
YEAR

INSULATION SPOT TEST GRAPH

5. Connect the megohmmeter leads for taking resistance measurements between A1 and ground (the motor frame).

6. Plot the insulation spot test measurements on the insulation spot test graph.

7. _____ Does the motor require service?

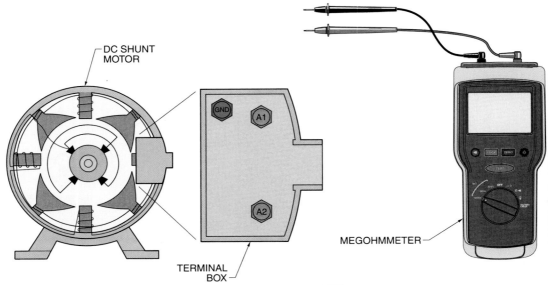

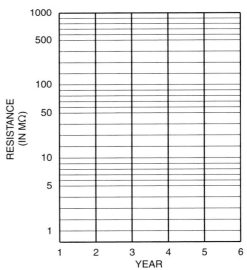

INSULATION SPOT TEST	
Test Date	**Resistance***
Jan, Year 1	500
Jul, Year 1	475
Jan, Year 2	450
Jul, Year 2	450
Jan, Year 3	425
Jul, Year 3	400
Jan, Year 4	400
Jul, Year 4	375
Jan, Year 5	375
Jul, Year 5	350
Jan, Year 6	350

* in MΩ

INSULATION SPOT TEST GRAPH

8. Connect the megohmmeter leads for taking resistance measurements between the first series winding terminal and the first shunt winding terminal.

9. Plot the insulation spot test measurements on the insulation spot test graph.

10. _____ During which period was the motor serviced?

11. _____ Did the servicing include replacing motor windings?

12. _____ Is the winding insulation in the refurbished motor performing adequately?

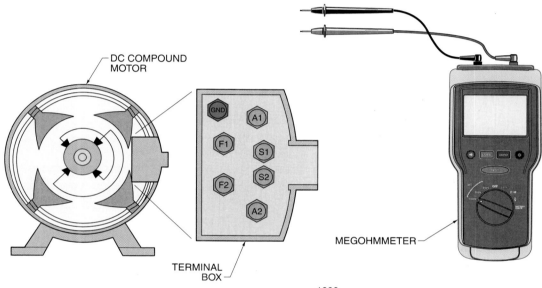

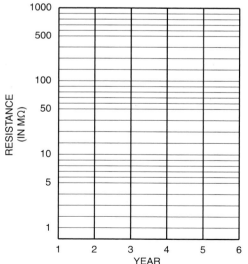

INSULATION SPOT TEST	
Test Date	**Resistance***
Jan, Year 1	400
Jul, Year 1	375
Jan, Year 2	300
Jul, Year 2	80
Jan, Year 3	10
Jul, Year 3	400
Jan, Year 4	350
Jul, Year 4	350
Jan, Year 5	325
Jul, Year 5	325
Jan, Year 6	300

* in MΩ

INSULATION SPOT TEST GRAPH

Insulation resistance also changes slightly over short periods (minutes) as voltage is applied to conductors. A dielectric absorption test measures the absorption characteristics of insulation by measuring resistance over a 10 min period. Measurements are recorded every 10 sec for the first minute and every minute thereafter. The readings can also be plotted on a graph to illustrate the changing resistance over time. The resistance of good insulation will increase continuously, appearing as an upward-sloping curve on the graph. The resistance of damaged or contaminated insulation will remain relatively constant, appearing as a flat curve. See Dielectric Absorption Test Graph.

Polarization index is a method of quantifying the minimum acceptable slope of the dielectric absorption test curve. The polarization index is calculated by dividing the values of the 10 min measurement by the 1 min measurement. A polarization index value that meets or exceeds the minimum value for the type of insulation tested indicates acceptable insulation. Insulation types are identified by class, such as Class A and Class F, and have different minimum polarization values, though most are between 1.5 and 2.0. See Polarization Index.

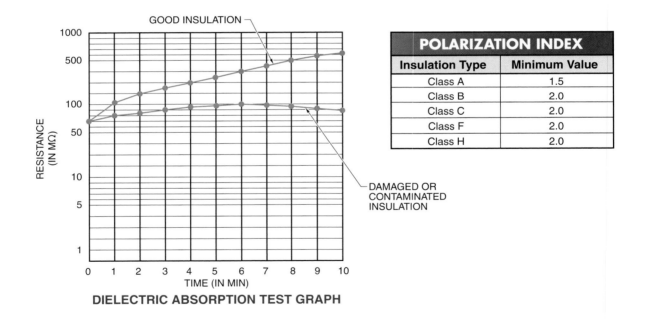

DIELECTRIC ABSORPTION TEST GRAPH

POLARIZATION INDEX	
Insulation Type	**Minimum Value**
Class A	1.5
Class B	2.0
Class C	2.0
Class F	2.0
Class H	2.0

It is usually not necessary to perform a dielectric absorption test and polarization index calculation on the insulation of every conductor in a motor or other load. Only one conductor with poor insulation will require servicing the load. A quicker insulation spot test on every conductor will identify the weakest insulation, which may then be further tested with a dielectric absorption test.

Megohmmeter measurements are taken on a 3ϕ motor from winding to winding and from each winding to ground. A dielectric absorption test is performed on the conductor with the lowest resistance measurement.

1. Plot the insulation test measurements on the dielectric absorption test graph.

2. _____ What is the polarization index value?

3. _____ Is the insulation resistance acceptable for Class B insulation?

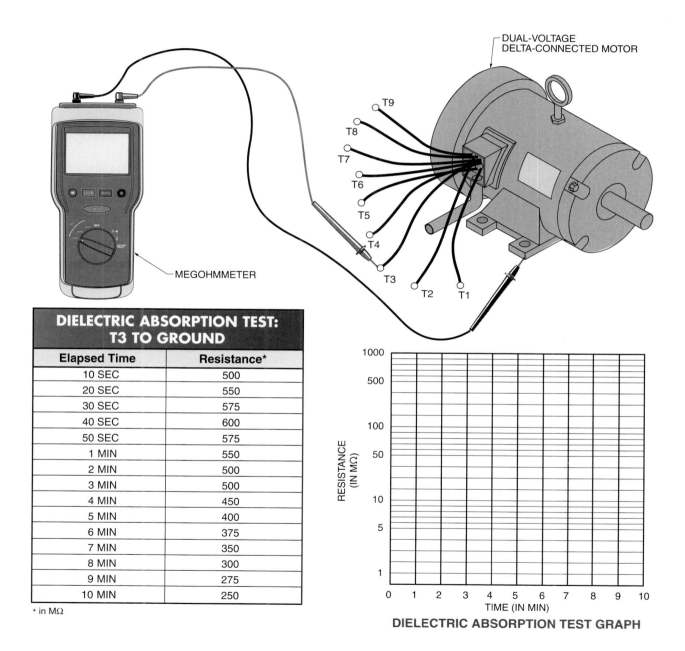

DIELECTRIC ABSORPTION TEST: T3 TO GROUND	
Elapsed Time	**Resistance***
10 SEC	500
20 SEC	550
30 SEC	575
40 SEC	600
50 SEC	575
1 MIN	550
2 MIN	500
3 MIN	500
4 MIN	450
5 MIN	400
6 MIN	375
7 MIN	350
8 MIN	300
9 MIN	275
10 MIN	250

* in MΩ

DIELECTRIC ABSORPTION TEST GRAPH

Megohmmeter measurements are taken on a heating element between ground and each wire leading to the heating element. A dielectric absorption test is performed on the conductor with the lowest resistance measurement.

4. Plot the insulation test measurements on the dielectric absorption test graph.

5. _____ What is the polarization index value?

6. _____ Is the insulation resistance acceptable for Class A insulation?

7. _____ Is the insulation resistance acceptable for Class F insulation?

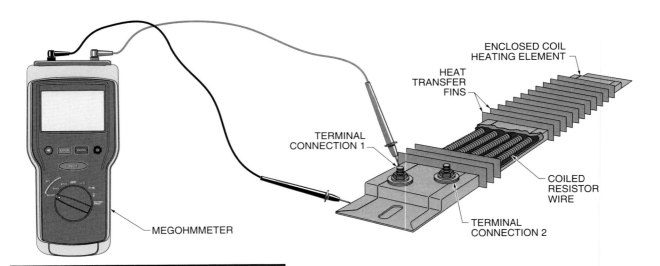

DIELECTRIC ABSORPTION TEST: TERMINAL 1 TO GROUND	
Elapsed Time	**Resistance***
10 SEC	100
20 SEC	150
30 SEC	175
40 SEC	175
50 SEC	200
1 MIN	200
2 MIN	225
3 MIN	275
4 MIN	300
5 MIN	325
6 MIN	325
7 MIN	350
8 MIN	350
9 MIN	325
10 MIN	350

* in MΩ

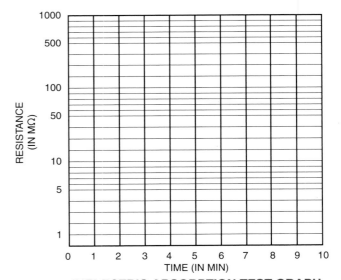

DIELECTRIC ABSORPTION TEST GRAPH

Insulation Step Voltage Testing

An insulation step voltage test reveals damaged or deteriorated areas on insulation by intentionally stressing the insulation to the point where weakened insulation fails. This test is only destructive when the insulation is already weakened to the point where it should be serviced anyway. The test should not damage insulation that is in good condition. This test is similar to hipot testing except that the displayed measurement is in ohms rather than amps.

During an insulation step voltage test, the resistance is measured as the voltage is increased incrementally from 1000 V to 5000 V. The resistance of good insulation remains relatively constant throughout the test. The resistance of damaged or deteriorated insulation decreases substantially as the voltage increases. See Insulation Step Voltage Test.

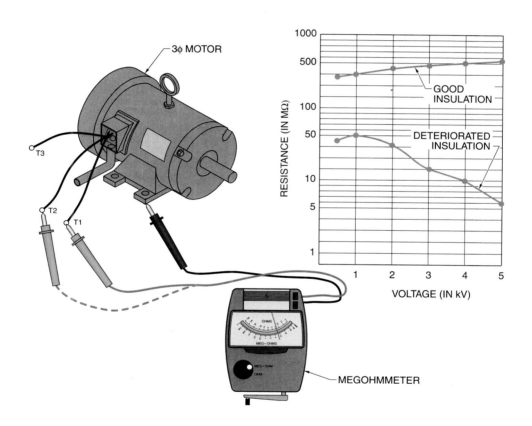

INSULATION STEP VOLTAGE TEST

Megohmmeter measurements are taken on a 3φ motor from winding to winding and from each winding to ground. An insulation step voltage test is performed on the conductors with the lowest resistance measurement.

1. Plot the insulation test measurements of T2 on the insulation step voltage test graph.

2. _____ Is the insulation resistance acceptable?

3. Plot the insulation test measurements of T3 on the insulation step voltage test graph.

4. _____ Is the insulation resistance acceptable?

5. _____ What can be done to repair the motor?

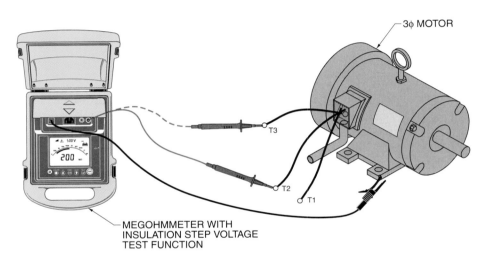

3φ MOTOR

MEGOHMMETER WITH
INSULATION STEP VOLTAGE
TEST FUNCTION

INSULATION STEP VOLTAGE TEST		
Voltage*	Resistance of T2 to Ground†	Resistance of T3 to Ground†
500	200	175
1000	200	200
1500	225	200
2000	225	150
2500	220	95
3000	220	80
3500	210	50
4000	200	40
4500	195	25
5000	190	10

* in V
† in MΩ

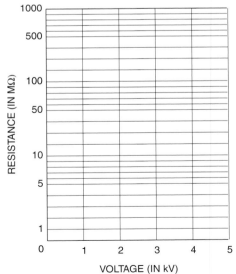

**INSULATION STEP
VOLTAGE TEST GRAPH**

Name_____ Date_____

_____ **1.** What unit is used to measure conductivity?

T   F   **2.** Resolution is the percentage of accuracy in an instrument.

T   F   **3.** Conductivity is the opposite of capacitance.

_____ **4.** What is the most common unit for measuring pressure?

_____ **5.** What is the accuracy (in psi) of a gauge that reads 80 psi and lists its accuracy as ±0.5% of reading?

_____ **6.** Which heat energy unit is defined as the amount of heat needed to raise the temperature of 1 lb of water by 1°F?

T   F   **7.** Pressure is force per unit of area.

_____ **8.** Which test instrument would most likely be used when determining if a corroded pipe wall is still structurally sound?

_____ **9.** Which two characteristics of a gas affect its relative humidity?

_____ **10.** What type of test instrument is used to check the accuracy of standard test instruments?

_____ **11.** Which has a higher thermal conductivity, water or gold?

T   F   **12.** The amount of uncertainty in a reading remains constant for any reading on an instrument that lists its accuracy as a percentage of full face value.

_____ **13.** What is gauge pressure at 25.3 psia?

T   F   **14.** Moisture test instruments measure the conductivity of the air.

_____ **15.** On a pressure gauge that measures up to 120 psi, does a reading of 100 psi ±0.5% of reading have a higher degree of accuracy than a reading of 50 psi ±0.5% of full face value?

T   F   **16.** Relative humidity is the amount of moisture in the air compared to the amount of moisture required to saturate the air.

_____ **17.** What type of energy does a noncontact temperature probe measure?

_____ **18.** Which type of gas leak detector uses a dye that is added to a refrigerant system?

_____ **19.** Which indicates a larger change in heat energy, one Celsius degree or one Fahrenheit degree?

_____ **20.** Which type of gas leak detector listens for the sounds created by a leak?

21. What types of changes can occur to substances in processes?

22. Give at least six examples of process variables that are measured with instrumentation.

23. Explain the difference between absolute pressure and gauge pressure, including how each measures vacuum.

24. How can a soapy solution be used to detect refrigerant or other gas leaks?

25. What are the major advantages and disadvantages of in-line and noncontact flowmeters?

Name _____   Date _____

Temperature Measurement Attachment

Test instruments are usually used to measure electrical quantities such as voltage, current, and resistance. Specialized instruments are often used to measure physical quantities such as speed, weight, and flow, and operational conditions such as temperature, humidity, and light. However, standard multimeters can also be used to take specialized measurements when used with sensor attachments that plug into the multimeter jacks. Attachments may be designed for digital or analog multimeters.

The readings are typically in mV or mA per unit being measured by the attachment, such as degrees, psi, or m/s. Careful attention must be paid to the output of each attachment so that the multimeter is set, connected, and read correctly. For example, an air velocity measurement attachment may output 1 mA/fps. The multimeter must be set to measure mA and the attachment must be connected to the mA and common jacks. The resulting measurement will display as mA, but is equivalent to fps (feet per second) of air velocity. See Multimeter Air Velocity Measurement Attachment.

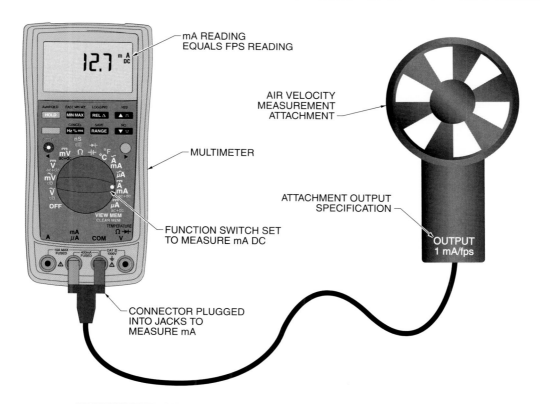

MULTIMETER AIR VELOCITY MEASUREMENT ATTACHMENT

A temperature measurement attachment is added to an analog multimeter for taking temperature readings.

1. Set (draw in) the multimeter selector switch to match the attachment's output.

2. Connect the attachment test leads to the multimeter to match the attachment's output.

3. _____ What is the temperature measurement of Reading 1?

4. _____ What is the temperature measurement of Reading 2?

The range switch is moved to the 10 V/1 V position and the red attachment test lead is moved to the +1 V jack.

5. _____ What is the temperature measurement of Reading 1?

6. _____ What is the temperature measurement of Reading 2?

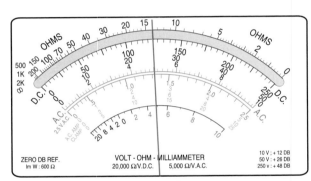

READING 1

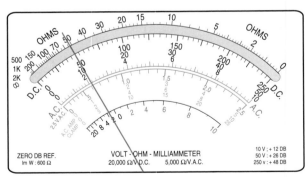

READING 2

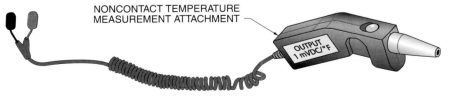

NONCONTACT TEMPERATURE
MEASUREMENT ATTACHMENT

OUTPUT
1 mVDC/°F

Instrumentation may call for measuring electrical current, which may be as small as a few milliamps on a process control loop, or several hundred amps on main power lines. Standard multimeters can only measure currents less than 10 A, so current attachments are used with multimeters to increase the current measuring range to hundreds or thousands of amps.

For measuring high current, a flexible current sensor loop is used to extend the current measuring range of a standard multimeter. The current measuring sensor reduces the actual circuit current proportionally (100:1, 1000:1, etc.) to a lower current that is sent to the meter attachment module. The meter attachment module converts the current signal to a proportional voltage signal that is then delivered to the meter. See Flexible Current Measuring Loop Attachment. Various models are available for taking current measurements from 0.5 A to 10,000 A. See Flexible Current Measuring Loop Models.

FLEXIBLE LOOP

CONTROL MODULE WITH OUTPUT SPECIFICATIONS

MULTIMETER CONNECTOR

**FLEXIBLE CURRENT
MEASURING LOOP ATTACHMENT**

FLEXIBLE CURRENT MEASURING LOOP MODELS			
Model	**Current Range***	**Output Signal†**	**Ratio**
A	0.5 to 30	100	10:1
B	0.5 to 200	1	1000:1
C	0.5 to 300	10 1	100:1 1000:1
D	0.5 to 30 0.5 to 300 ‡	100 10	10:1 100:1
E	0.5 to 500	1	1000:1
F	0.5 to 50 0.5 to 500 ‡	10 1	100:1 1000:1
G	0.5 to 1000	1	1000:1
H	0.5 to 100 0.5 to 1000 ‡	10 1	100:1 1000:1
I	0.5 to 3000	1	1000:1
J	0.5 to 300 0.5 to 3000 ‡	10 1	100:1 1000:1
K	0.5 to 6000	0.1	10,000:1
L	0.5 to 600 0.5 to 6000 ‡	1 0.1	1000:1 10,000:1
M	0.5 to 10,000	0.1	10,000:1
N	0.5 to 1000 0.5 to 10,000 ‡	1 0.1	1000:1 10,000:1

* in A
† in mVAC/A
‡ dual-range models

Various current loop attachments are used with the multimeter to take high current measurements.

1. Connect the current attachment connector to the proper input jacks of the multimeter.

2. _____ What is the Model A attachment's output signal to the multimeter?

3. _____ If a Model A attachment is around a conductor carrying 15 A, what will the multimeter measure?

4. _____ If a Model A attachment is around a conductor carrying 30 A, what will the multimeter measure?

5. _____ If a Model K attachment is around a conductor carrying 6000 A, what will the multimeter measure?

6. _____ What is the maximum output signal specified for any of the current loop models?

7. Set (draw in) the function and range switches so that any current loop model can be used.

The multimeter range switch is set to 2.5 V for taking measurements.

8. _____ If Reading 1 is displayed when a Model I attachment is around a conductor, what is the actual current?

9. _____ If Reading 1 is displayed when a Model A attachment is around a conductor, what is the actual current?

10. _____ If Reading 2 is displayed when a Model I attachment is around a conductor, what is the actual current?

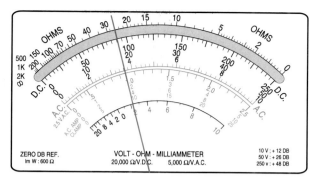

READING 1

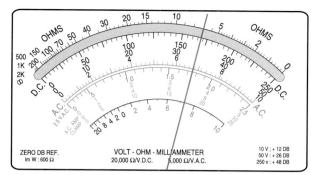

READING 2

HIGH CURRENT
CONDUCTOR

11. Connect the current attachment connector to the proper input jacks of the multimeter.

12. Set (draw in) the function switch so that any current loop model can be used.

13. _____ If Reading 1 is displayed when a Model G attachment is around a conductor, what is the actual current?

14. _____ If Reading 1 is displayed when a Model M attachment is around a conductor, what is the actual current?

15. _____ If Reading 2 is displayed when a Model B attachment is around a conductor, what is the actual current?

16. _____ If Reading 2 is displayed when a Model H attachment (set to 100:1) is around a conductor, what is the actual current?

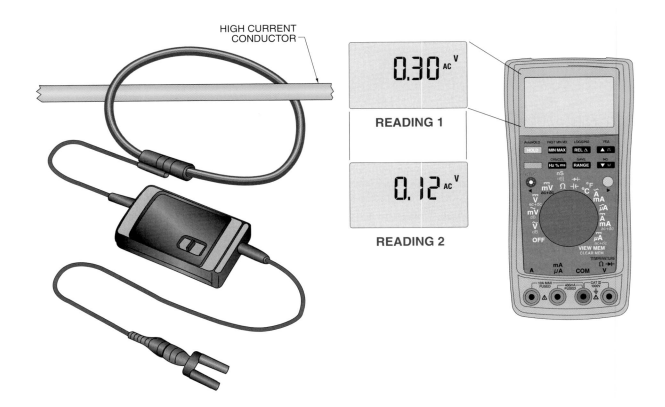

Thermocouples

A thermocouple is a temperature sensor that produces a voltage output nearly proportional to the temperature. A thermocouple consists of two dissimilar metals joined at one end. Current flows through the thermocouple when the welded joint is heated. A small voltage output, typically only a few millivolts DC, is measured at the open junction. Higher voltage indicates a higher temperature at the heated end. See Thermocouples.

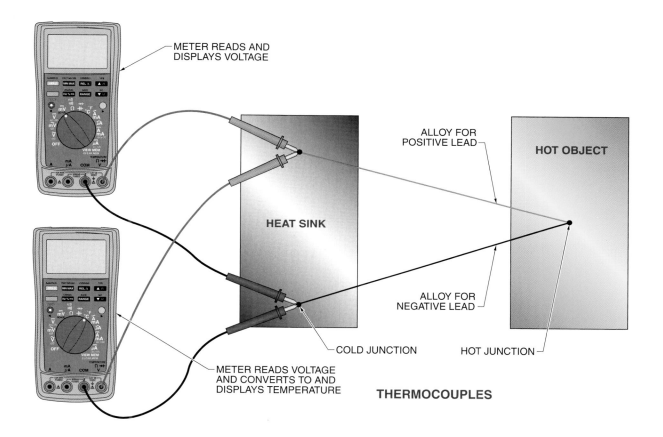

THERMOCOUPLES

The heated end of a thermocouple is known as the hot junction. The other end is known as the cold junction. For the thermocouple's output voltage to be converted into a true temperature measurement, some compensation for temperature differences between the two junctions must be made. A temperature compensator ensures that the correct thermocouple output is delivered to the measuring point, regardless of any changing ambient temperatures. The temperature compensator is placed between the thermocouple and the measuring controller or meter.

Thermocouples are the most common type of temperature sensor used in industrial heating applications. One of the reasons thermocouples are so common is that they have the highest temperature-measuring capability of any sensor type. Thermocouples can measure temperatures up to 3000°F. They are also low in cost, rugged, small, have a fast response time, and are available in a wide range of sizes.

Thermocouples come in a variety of assemblies, are suited to a variety of temperature ranges, and are available for a variety of operating conditions. Standard color-coded leads are used to indicate the thermocouple type and which lead is positive and which lead is negative. See Thermocouple Color Codes.

THERMOCOUPLE COLOR CODES						
ANSI	**Alloy Combination**		**Thermocouple Grade**		**Extension Grade**	
	Positive Lead	**Negative Lead**	**Color Code***	**Maximum Temperature Range**	**Color Code***	**Maximum Temperature Range**
J	IRON Fe (magnetic)	COPPER-NICKEL Cu-Ni (CONSTANTAN)		–210°C to 1200°C –346°F to 2193°F		0°C to 200°C 32°F to 392°F
K	NICKEL-CHROMIUM Ni-Cr (CHROMEL)	NICKEL-ALUMINUM Ni-Al (magnetic) (ALUMEL)		–270°C to 1372°C –454°F to 2501°F		0°C to 200°C 32°F to 212°F
T	COPPER Cu	COPPER-NICKEL Cu-Ni (CONSTANTAN)		–270°C to 400°C –454°F to 752°F		–60°C to 100°C –76°F to 212°F
E	NICKEL-CHROMIUM Ni-Cr (CHROMEL)	COPPER-NICKEL Cu-Ni (CONSTANTAN)		–270°C to 1000°C –454°F to 1832°F		0°C to 200°C 32°F to 392°F
N	NICKEL-CHROMIUM-SILICON Ni-Cr-Si	NICKEL-SILICON-MAGNESIUM Ni-Si-Mg		–270°C to 1300°C –454°F to 1832°F		0°C to 200°C 32°F to 392°F
R	PLATINUM-13% RHODIUM Pt-13% Rh	PLATINUM Pt	NOT ESTABLISHED	–50°C to 1768°C –58°F to 3214°F		0°C to 150°C 32°F to 392°F
S	PLATINUM-10% RHODIUM Pt-10% Rh	PLATINUM Pt	NOT ESTABLISHED	–50°C to 1768°C –58°F to 3214°F		0°C to 150°C 32°F to 300°F
B	PLATINUM-30% RHODIUM Pt-30% Rh	PLATINUM-6% RHODIUM Pt-6% Rh	NOT ESTABLISHED	50°C to 1820°C 122°F to 3308°F		0°C to 100°C 32°F to 212°F

* U.S. and Canadian color codes

The various types of thermocouples are different because of the pairs of metals that are used. Specific alloys may be better suited to certain operating environments, such as immersion in corrosive liquids. The different alloy combinations also produce thermocouples for different temperature ranges. Some thermocouples may have larger temperature ranges and others may have shorter ranges but are more accurate. A more steeply sloped plot on a temperature-voltage graph means that the thermocouple has a greater voltage change for each degree change in temperature. This indicates that the thermocouple allows more precise temperature measurement. See Thermocouple Temperature-Voltage Relationships.

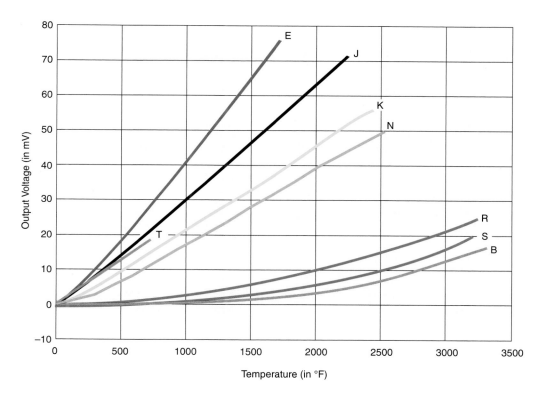

THERMOCOUPLE TEMPERATURE-VOLTAGE RELATIONSHIPS

A type J thermocouple monitors the temperature of a process fluid. A multimeter is used to read the voltage output directly.

1. Set (draw in) the function switch of the multimeter to test the output of the thermocouple.

2. Connect the multimeter test leads to the correct jacks on the multimeter.

3. Connect the multimeter test leads to test the output of the thermocouple at the temperature compensator.

4. _____ If the multimeter reads 14 mV, what is the approximate temperature of the fluid?

5. _____ If the multimeter reads 27 mV, what is the approximate temperature of the fluid?

6. _____ If the multimeter reads 30 mV, what is the approximate temperature of the fluid?

The thermocouple is changed to a type E, which should be more precise in the temperature range of the process fluid.

7. _____ If the fluid is 500°F, what should the multimeter approximately read?

8. _____ If the fluid is 1000°F, what should the multimeter approximately read?

9. _____ If the multimeter reads 30 mV, what is the approximate temperature of the fluid?

10. _____ If the multimeter reads 24 mV, what is the approximate temperature of the fluid?

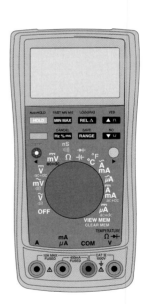

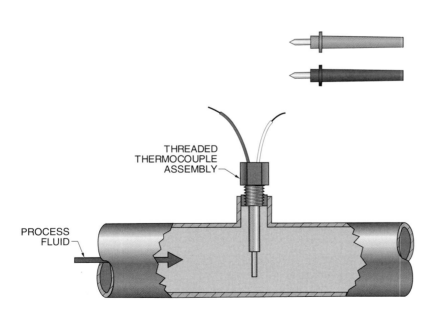

THREADED
THERMOCOUPLE
ASSEMBLY

PROCESS
FLUID

Application 8-4

Process Control Signals

Standard control signals are used to represent possible analog values of a process variable. Control signals may be electrical, using variable voltage or variable current, or pneumatic, using variable air pressure. For example, if a temperature in a process tank varies from 70°F to 170°F, a temperature sensor can be used to send a signal to a controller that reads the temperature and converts it to a standard signal between 4 mA DC and 20 mA DC. A controller with a variable current output would be calibrated (adjusted) to produce a 4 mA signal at 70°F and a 20 mA signal at 170°F. Thus, when the temperature controller reads 120°F (halfway between 70°F and 170°F), the controller would output a 12 mA signal (halfway between 4 mA and 20 mA) and any actuators controlled by the signal would be halfway between their two extreme states (such as halfway between fully open and fully closed).

A system regulates the temperature of a process fluid between 50°F and 150°F by cooling or heating the fluid as needed. The cooling and heating units each have a valve that opens an amount proportional to the amount of cooling or heating needed. A thermocouple measures the fluid's temperature and a temperature controller outputs the corresponding control signal that operates the valves. The multimeter measures the temperature controller's current output.

1. Set (draw in) the function switch on the multimeter for measuring the control signal.

2. Connect the multimeter test leads to the correct jacks on the multimeter.

3. _____ If the fluid temperature is 100°F, what should the multimeter display?

4. _____ If the fluid temperature is 100°F, what is the condition of the cooling valve?

5. _____ If the fluid temperature is 100°F, what is the condition of the steam valve?

6. _____ If the fluid temperature is 75°F, what should the multimeter display?

7. _____ If the fluid temperature is 75°F, what is the condition of the cooling valve?

8. _____ If the fluid temperature is 75°F, what is the condition of the steam valve?

9. _____ If the fluid temperature is 125°F, what should the multimeter display?

10. _____ If the fluid temperature is 125°F, what is the condition of the cooling valve?

11. _____ If the fluid temperature is 125°F, what is the condition of the steam valve?

12. _____ If the fluid temperature is 50°F, what should the multimeter display?

13. _____ If the fluid temperature is 50°F, what is the condition of the cooling valve?

14. _____ If the fluid temperature is 50°F, what is the condition of the steam valve?

15. _____ What is the ideal temperature that this system is trying to maintain?

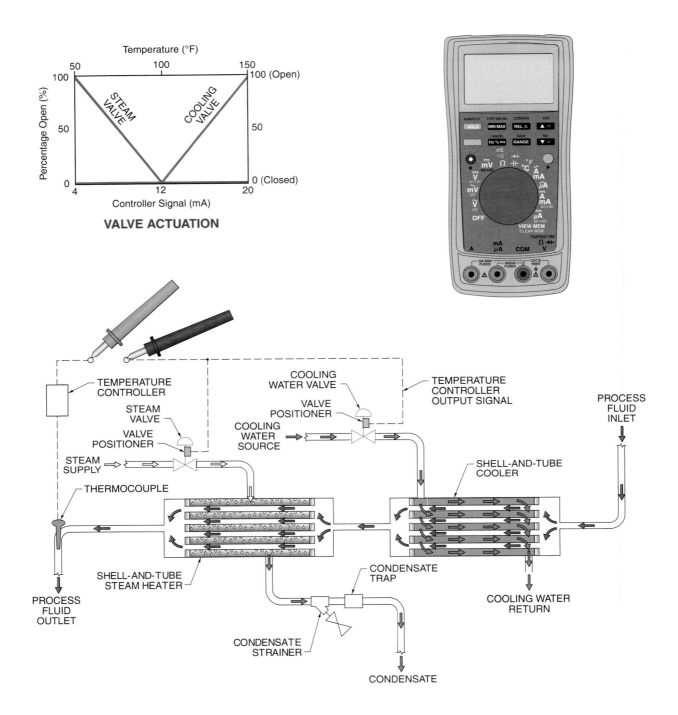

VALVE ACTUATION

Application 8-5 *Motor Drive Control Signals*

Electric motor drives are ideal for controlling pump and fan motors, because the speed of the motor can be varied to control the flow rate. The motor is connected to the drive, which is connected to the incoming supply voltage. Control signals into the drive can be digital signals to turn the motor ON or OFF, or analog signals to control motor speed. See Motor Drive Control Signal Inputs and Motor Drive Input Examples.

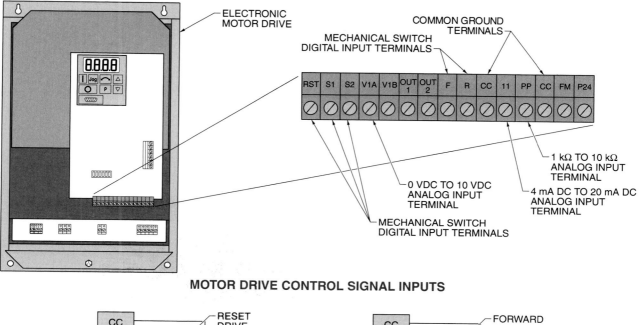

MOTOR DRIVE CONTROL SIGNAL INPUTS

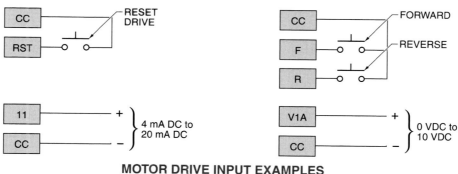

MOTOR DRIVE INPUT EXAMPLES

The digital signals can be from mechanical switches, such as pushbuttons, pressure switches, or temperature switches, or from the output contacts of relays or PLCs. Relays, especially solid-state relays, are used to input control signals from electronic circuits.

The analog signals control the speed of the motor as a percent of total motor speed. For example, an analog signal of 0 VDC to 10 VDC would increment the motor speed by 10% of motor maximum speed for every 1 VDC of input signal. The maximum speed of a motor is the nameplate-rated speed (in rpm) at the nameplate-rated frequency (in Hz). Changing the frequency changes the speed of the motor.

An electric motor drive can be set to output a maximum frequency lower or higher than the motor's nameplate-rated frequency. Also, an analog signal input of 0 VDC to 10 VDC into a drive would set the frequency at 10% of the drive's (not necessarily the motor's) maximum frequency for every 1 VDC input. Therefore, the analog signal is incrementing the drive frequency up to a set maximum, which may or may not match the motor's maximum frequency. The drive output frequency can be calculated by applying the following formula:

$$f_{OUT} = \frac{S - S_{MIN}}{S_{MAX} - S_{MIN}} \times f_D$$

where

f_{OUT} = frequency output of drive (in Hz)

S = control signal (in V or mA)

S_{MAX} = maximum control signal (in V or mA)

S_{MIN} = minimum control signal (in V or mA)

f_D = maximum frequency setting of drive (in Hz)

For example, a 6 VDC signal (out of 0 VDC to 10 VDC) into a motor drive set for a maximum frequency of 30 Hz results in a frequency output of 18 Hz.

$$f_{OUT} = \frac{S - S_{MIN}}{S_{MAX} - S_{MIN}} \times f_D$$

$$f_{OUT} = \frac{6 - 0}{10 - 0} \times 30$$

$$f_{OUT} = 0.6 \times 30$$

$$f_{OUT} = \textbf{18 Hz}$$

The output frequency of the motor drive controls the speed of the motor. The speed of the motor is proportional to the drive frequency and can be calculated by applying the following formula:

$$n = \frac{f_{OUT}}{f_M} \times n_M$$

where

n = motor speed (in rpm)

f_{OUT} = frequency output of drive (in Hz)

f_M = nameplate-rated frequency of motor (in Hz)

n_M = nameplate-rated speed of motor (in rpm)

For example, a motor drive outputs 18 Hz to a motor with a nameplate rated speed of 3550 rpm at 60 Hz. The motor speed is 1065 rpm.

$$n = \frac{f_{OUT}}{f_M} \times n_M$$

$$n = \frac{18}{60} \times 3550$$

$$n = 0.3 \times 3550$$

$$n = \textbf{1065 rpm}$$

A motor with a nameplate-rated speed of 1740 rpm at 60 Hz is controlled by a motor drive. The motor drive is set for a maximum frequency output of 60 Hz and receives control signals of 0 VDC to 10 VDC.

1. Set (draw in) the function switch for measuring the analog control signal into the drive.

2. Connect the multimeter test leads to the correct jacks for measuring the control signal at the drive.

3. Connect the test leads to measure the control signal circuit of the drive.

4. _____ What is the output frequency when the multimeter reads 4 VDC?

5. _____ What is the motor speed when the multimeter reads 4 VDC?

The motor drive is reset to a maximum output frequency of 66 Hz.

6. _____ What is the output frequency when the multimeter reads 4 VDC?

7. _____ What is the motor speed when the multimeter reads 4 VDC?

8. _____ What is the output frequency when the multimeter reads 6 VDC?

9. _____ What is the motor speed when the multimeter reads 6 VDC?

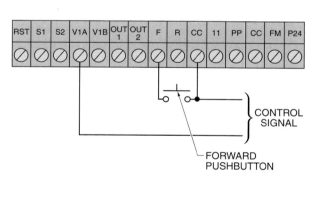

A motor with a nameplate-rated speed of 1740 rpm at 60 Hz is controlled by a motor drive. The motor drive is set for a maximum frequency output of 60 Hz and receives control signals of 4 mA DC to 20 mA DC.

10. Set (draw in) the function switch for measuring the analog control signal into the drive.

11. Connect the multimeter test leads to the correct jacks for measuring the control signal at the drive.

12. Connect the test leads to measure the control signal circuit of the drive.

13. _____ What is the output frequency when the multimeter reads 6 mA?

14. _____ What is the motor speed when the multimeter reads 6 mA?

15. _____ What is the output frequency when the multimeter reads 18 mA?

16. _____ What is the motor speed when the multimeter reads 18 mA?

The motor drive is reset to a maximum output frequency of 50 Hz.

17. _____ What is the output frequency when the multimeter reads 10 mA?

18. _____ What is the motor speed when the multimeter reads 10 mA?

19. _____ What is the output frequency when the multimeter reads 15 mA?

20. _____ What is the motor speed when the multimeter reads 15 mA?

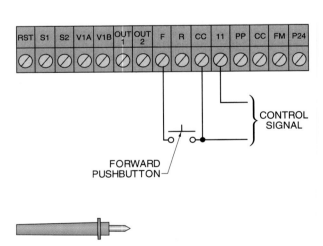

Name_____ Date_____

_____ 1. Which unit is used to measure the amount of light emitted by a lamp?

_____ 2. Which type of tachometer uses a flashing light to measure speed?

T   F   3. Recommended light levels in buildings vary by industry, venue, and event.

_____ 4. Which type of test instrument is used to measure rotational speed?

_____ 5. Which types of tachometers use a reflective area or piece of reflective tape on the rotating object to count revolutions?

_____ 6. Where is the transmitter unit of a branch circuit identifier plugged in?

T   F   7. All power must be turned OFF to a component before testing it with a micro-ohmmeter.

_____ 8. Which type of test instrument accurately measures very small resistances?

T   F   9. Rotational speed can be measured in feet per minute (fpm).

_____ 10. Which unit of illumination is equal to one lumen per square foot?

_____ 11. How does a gas detector indicate when the levels of a gas become unsafe?

_____ 12. Which part of a vibration meter is attached to the object being measured?

T   F   13. Laser and photo tachometers usually have wider measurement ranges than contact tachometers.

_____ 14. What is the illumination (in fc) of 50 lumens on 5 sq ft of area?

_____ 15. Which unit of illumination is equal to 1 lumen per square meter?

T   F   16. Micro-ohmmeters can be used to check mechanical connections such as welds and fastened joints.

T   F   17. Motors must be locked out and tagged out before measurements are taken with a tachometer.

_____ 18. What is the illumination of 280 fc converted to lux?

_____ 19. Is the wavelength of infrared energy longer or shorter than visible light?

_____ 20. Which color are the shortest wavelengths of visible light?

21. Why are noncontact tachometers safer than contact tachometers?

22. What are some of the potential dangers of a confined space?

23. How does the wavelength of energy from light sources affect the appearance of colors?

24. How is a branch circuit identifier used to identify the correct circuit breaker for a particular receptacle?

25. What is the difference between how light is measured for indoor and outdoor applications?

Name _____   Date _____

Micro-Ohmmeter Measurements

Often, circuit devices such as contacts, switches, and splices are assumed to have a resistance so low that their effect on the circuit is negligible. See Neglecting Contact Resistance. However, very low resistance electrical connections, and even the conductors, will cause a voltage drop in a current-carrying circuit. The voltage drop results in power loss in the form of excess heat. If the circuit is very sensitive to voltage changes or excess heat, the small voltage drops can affect circuit operation.

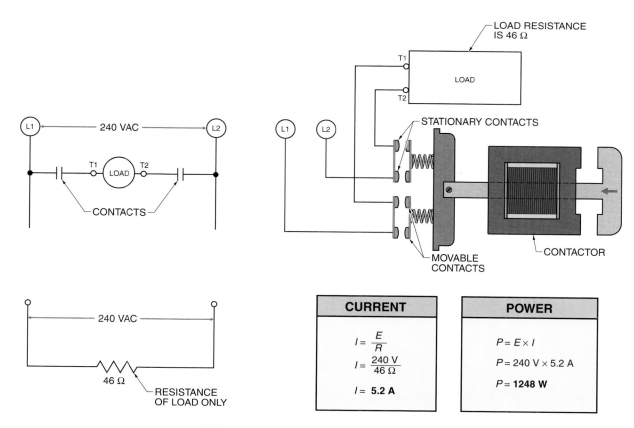

CURRENT	POWER
$I = \dfrac{E}{R}$	$P = E \times I$
$I = \dfrac{240\text{ V}}{46\ \Omega}$	$P = 240\text{ V} \times 5.2\text{ A}$
$I = $ **5.2 A**	$P = $ **1248 W**

NEGLECTING CONTACT RESISTANCE

Micro-ohmmeters are used to measure very low resistances of less than 1 Ω, and some measure as low as 0.000001 Ω. Measuring the additional resistances of connections such as conductors, contacts, switches, and splices builds a more complete model of an actual circuit. Total circuit resistance, voltage drops, and power loss can be calculated with Ohm's law, the power formula, and the laws of series and parallel circuits. See Including Contact Resistance.

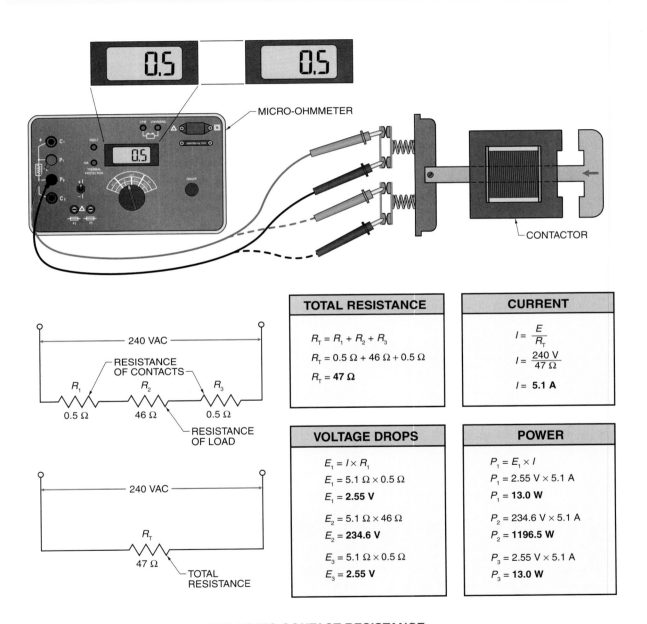

TOTAL RESISTANCE
$R_T = R_1 + R_2 + R_3$
$R_T = 0.5\ \Omega + 46\ \Omega + 0.5\ \Omega$
$R_T = \textbf{47}\ \boldsymbol{\Omega}$

CURRENT
$I = \dfrac{E}{R_T}$
$I = \dfrac{240\ \text{V}}{47\ \Omega}$
$I = \textbf{5.1 A}$

VOLTAGE DROPS
$E_1 = I \times R_1$
$E_1 = 5.1\ \Omega \times 0.5\ \Omega$
$E_1 = \textbf{2.55 V}$
$E_2 = 5.1\ \Omega \times 46\ \Omega$
$E_2 = \textbf{234.6 V}$
$E_3 = 5.1\ \Omega \times 0.5\ \Omega$
$E_3 = \textbf{2.55 V}$

POWER
$P_1 = E_1 \times I$
$P_1 = 2.55\ \text{V} \times 5.1\ \text{A}$
$P_1 = \textbf{13.0 W}$
$P_2 = 234.6\ \text{V} \times 5.1\ \text{A}$
$P_2 = \textbf{1196.5 W}$
$P_3 = 2.55\ \text{V} \times 5.1\ \text{A}$
$P_3 = \textbf{13.0 W}$

INCLUDING CONTACT RESISTANCE

A 240 VAC circuit has a contactor switching a load that has 158 Ω of resistance. The micro-ohmmeter measures 100 mΩ for each set of contacts.

1. _____ What is the total circuit resistance?

2. _____ What is the current (in A) through the circuit?

3. _____ What is the voltage drop (in V) across each contact?

4. _____ How much power (in mW) is produced at each contact?

5. _____ How much power (in mW) is produced at the load?

6. _____ If the resistance of the contacts is neglected, how much power is produced at the load?

A 240 VAC circuit has a contactor switching a load that has 67 Ω of resistance. The micro-ohmmeter measures 50 mΩ for each set of contacts.

7. _____ What is the total circuit resistance?

8. _____ What is the current (in A) through the circuit?

9. _____ What is the voltage drop (in V) across each contact?

10. _____ How much power (in mW) is produced at each contact?

11. _____ How much power (in mW) is produced at the load?

12. _____ If the resistance of the contacts is neglected, how much power is produced at the load?

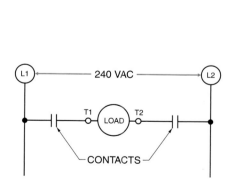

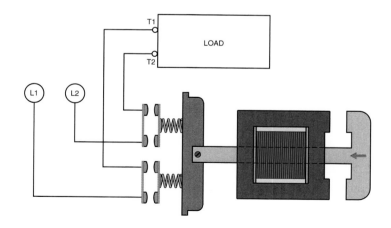

The speed of a motor is measured using a contact tachometer, photo tachometer, or laser tachometer. Torque is the rotational force at the shaft of the motor. A motor under greater load requires more torque to turn the shaft. Horsepower is the combination of speed and torque together.

Horsepower, torque, and speed are related such that when any two are known, the third quantity can be determined. For example, if speed and horsepower are known, torque can be determined. If speed and torque are known, horsepower can be determined, and if torque and horsepower are known, speed can be determined.

Conversion charts can give quick approximations of the unknown value when the other two are known. Connecting points on any two scales defines a line which approximately indicates the third value where the line intersects the third scale. See Horsepower-Torque-Speed Conversion.

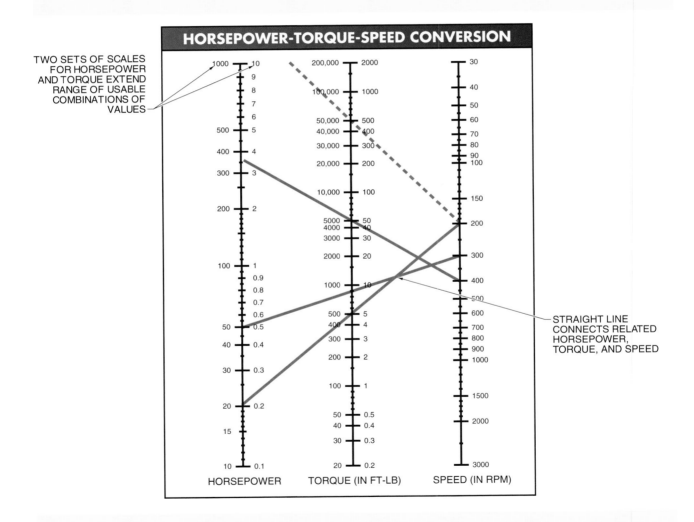

The horsepower and torque scales cover wider ranges by including two sets of numbers. The numbers on the left side of the scales are 100 times larger than the numbers on the right side. When using the conversion chart, the same sides of the horsepower and torque scales must be used together. The number values and the line connecting them determine which side to use.

For example, the line connecting 200 rpm with 500 ft-lb (on the right side of the torque scale) extends off the conversion chart and horsepower cannot be determined. However, connecting 200 rpm with 500 ft-lb on the left side of the torque scale yields a usable line. The result is approximately 20 HP. The left side of the horsepower scale must be used because the left side of the torque scale was used.

Also, the conversion chart shows that 50 ft-lb of torque at 400 rpm produces approximately 3.6 HP and that a ½ HP motor turning at 300 rpm produces approximately 9 ft-lb of torque.

Conversion charts are useful for approximating values such as torque, but calculating the torque gives a more accurate value. Torque is found by applying the following formula:

$$T = \frac{HP}{n} \times 5252$$

where

T = torque (in ft-lb)

HP = power (in HP)

n = rotational speed (in rpm)

For example, what is the full-load torque of a 60 HP, 240 V, 3ϕ motor turning at 1725 rpm?

$$T = \frac{HP}{n} \times 5252$$

$$T = \frac{60}{1725} \times 5252$$

$$T = 0.0348 \times 5252$$

$$T = \textbf{182.7 ft-lb}$$

A laser tachometer measures 120 rpm on a 50 HP conveyor motor.

1. _____ What is the approximate torque (using the conversion chart)?

2. _____ What is the calculated torque?

3. _____ If the motor slows down to 100 rpm, what is the calculated torque?

4. _____ If the motor speeds up to 200 rpm, what is the calculated torque?

A laser tachometer measures 1700 rpm on a 2 HP conveyor motor.

5. _____ What is the approximate torque (using the conversion chart)?

6. _____ What is the calculated torque?

7. _____ If the motor speeds up to 2000 rpm, what is the calculated torque?

8. _____ If heavier boxes are placed on the conveyor, what might happen to the motor speed?

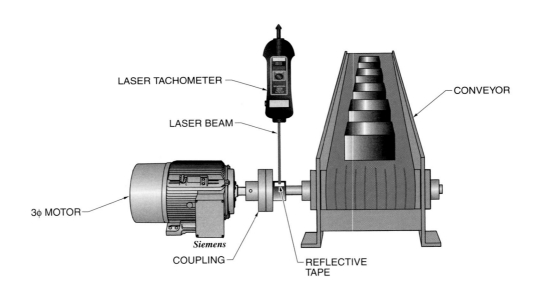

Carbon Monoxide Testing

A carbon monoxide (CO) probe is a multimeter attachment that measures the level of CO in an atmosphere. Carbon monoxide is a colorless and odorless toxic gas that can cause serious health problems or death by obstructing the transfer of oxygen in the bloodstream. Common sources of carbon monoxide emissions include malfunctioning or improperly installed furnaces or fireplaces, obstructed or undersized chimneys or flue exhausts, poorly maintained gas, oil, or kerosene appliances, and improperly ventilated combustion engines such as automobile engines, portable generators, and lawnmowers. See Carbon Monoxide (CO) Emission Sources.

CARBON MONOXIDE (CO) EMISSION SOURCES		
Appliance	**Fuel**	**Typical problems**
Gas furnaces; room heaters	Oil, natural gas, or LPG (liquefied petroleum gas)	1. Cracked heat exchanger 2. Not enough air to burn fuel properly 3. Defective/blocked flue 4. Poorly adjusted burner 5. Building not properly pressurized
Central heating furnaces	Coal or kerosene	1. Cracked heat exchanger 2. Not enough air to burn fuel properly 3. Defective grate
Room heaters; central heaters	Kerosene	1. Improper adjustment 2. Wrong fuel 3. Wrong wick or wick height 4. Not enough air to burn fuel 5. System not properly vented
Water heaters	Natural gas or LPG	1. Not enough air to burn fuel properly 2. Poorly adjusted burner 3. Misuse as a room heater 4. System not properly vented
Ranges; ovens	Natural gas or LPG	1. Not enough air to burn fuel properly 2. Poorly adjusted burner 3. Misuse as a room heater 4. System not properly vented
Stoves; fireplaces	Gas, wood, coal	1. Not enough air to burn fuel properly 2. Defective/blocked flue 3. Green or treated wood 4. Cracked heat exchanger 5. Cracked firebox

Symptoms of carbon monoxide poisoning include headaches, dizziness, sleepiness, and general weakness. High levels of carbon monoxide can be deadly. Personnel should be evacuated and carbon monoxide levels checked if anyone reports these symptoms. For maximum safety, carbon monoxide measurements should be taken whenever working on or around devices capable of producing carbon monoxide. Carbon monoxide levels are also measured when evaluating a confined space.

A carbon monoxide probe multimeter attachment measures the level of carbon monoxide in parts per million (ppm). Permissible carbon monoxide exposure levels are set by governing agencies. See Carbon Monoxide (CO) Level Standards.

CARBON MONOXIDE (CO) LEVEL STANDARDS	
0 to 1 PPM	Normal background levels
9 PPM	ASHRAE* limit for living areas
50 PPM	OSHA† enclosed space 8-hour average level
100 PPM	OSHA† exposure limit
200 PPM	Mild headache, fatigue, nausea, and dizziness
800 PPM	Dizziness, nausea, and convulsions; death within 2 to 3 hours

* American Society of Heating, Refrigeration, and Air-Conditioning Engineers, Inc.
† Occupational Safety and Health Administration

A service call requires that a warehouse be checked for carbon monoxide levels. The warehouse staff uses a variety of electric, propane-powered, and gas-powered vehicles to transport products.

1. Connect the carbon monoxide probe attachment to the correct multimeter jacks.

2. Set (draw in) the correct function switch position for taking carbon monoxide level measurements.

3. _____ If the multimeter measures levels between 10 mV and 85 mV at various warehouse locations, do the carbon monoxide levels meet the OSHA exposure limit for an 8 hr shift?

4. _____ If the multimeter measures levels between 10 mV and 85 mV at various warehouse locations, do the carbon monoxide levels meet the OSHA exposure limit for a 10 hr shift?

5. _____ If the multimeter measures levels between 10 mV and 125 mV, and an average level of 30 mV, at various warehouse locations, are the carbon monoxide levels above a dangerous level at any location within the warehouse?

6. _____ If the multimeter reads 25 mV and the relative mode is activated, what will the multimeter read?

7. _____ If the multimeter reads +30 mV REL, what is the actual level of carbon monoxide at that location?

8. _____ If the multimeter reads –8 mV REL, what is the actual level of carbon monoxide at that location?

9. _____ If the multimeter reads +52 mV MAX REL, what is the maximum level of carbon monoxide in the warehouse?

10. _____ If the multimeter reads –12 mV MIN REL, what is the minimum level of carbon monoxide in the warehouse?

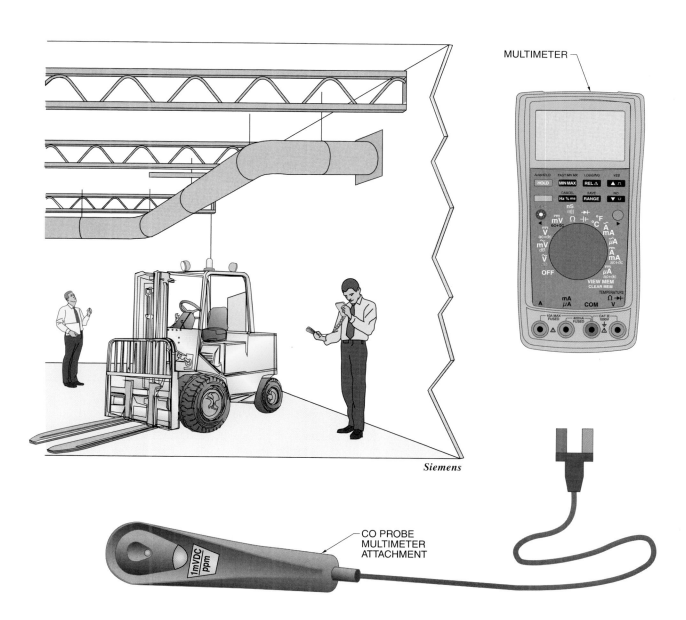

Siemens

MULTIMETER

CO PROBE
MULTIMETER
ATTACHMENT

                 Light Level Testing

The proper amount of light ensures that living and working conditions are safe and secure, and that environments are favorable for playing sports, displaying merchandise, or other activities. It is important to achieve good lighting in the right amount for the application. Electricians should know how to measure light levels and make recommendations for lighting types and levels in new installations, or additional lighting in existing systems.

Lamp (light bulb) manufacturers produce many different lamp types and sizes for various applications. Lamps are rated for their power consumption (in watts), light output (in lumens or lumens per watt), life expectancy (in hours), and features such as light color and energy efficiency. See Lamp Types.

LAMP TYPES					
Type	**Shape**	**Power Consumption***	**Light Output†**	**Average Rated Life‡**	**Features**
Mercury vapor		40 to 1000	50 to 60	24,000 +	Long life; low price
Metal-halide		35 to 1500	80 to 125	6000 to 20,000	Good color rendering; long life; high efficiency
High-pressure sodium		35 to 1000	80 to 150	15,000 to 24,000	Extremely long life; good lumen maintenance; high efficiency
Low-pressure sodium		18 to 180	190 to 200	18,000	Very long life; extremely high efficiency; poor color rendering
Fluorescent		4 to 215	55 to 100	6000 to 36,000	Wide choice of light colors; excellent color rendering; good uniformity; long life; very efficient
Incandescent		3 to 1500	15 to 25	500 to 8000	Easy to install; many shapes; low cost; instant start
Halogen		45 to 1500	18 to 22	2000 to 6000	Compact; high output; white light; easy to install

* in W
† in lm/W
‡ in hr

Light meters are used to measure the amount of light at any location. The amount of light is usually specified and measured in footcandles (fc) or lumens (lm). Charts list recommended light levels for various applications such as offices, classrooms, and sporting events. Most light level charts and specifications use footcandle units. See Recommended Light Levels.

RECOMMENDED LIGHT LEVELS

INTERIOR LIGHTING				EXTERIOR LIGHTING	
Area	**Light Level***	**Area**	**Light Level***	**Area**	**Light Level***

Assembly
  Rough, easy seeing
  Medium
  Fine
	
	30
	100
	500
		Machine shop
		  Rough bench
		  Medium bench
			
			50
			100
				Building
				  Light surface
				  Dark surface
					
					15
					50

Auditorium
  Exhibitions
	
	30
		Materials handling
		  Picking stock
		  Packing, labeling
			
			30
			50
				Loading/unloading area	20

Banks
  Lobby, general
  Writing areas
  Teller station
	
	50
	70
	150
		Offices
		  Regular office work
		  Accounting
		  Detailed work
			
			100
			150
			200
				Parking areas
				  Industrial
				  Shopping
					
					2
					5

Canning
  Cutting, sorting
	
	100
		Printing
		  Proofreading
		  Color inspecting
			
			150
			200
				Storage yards (active)	20

Clothing manufacturing
  Patternmaking
  Shops
	
	50
	100
		Schools
		  Auditoriums
		  Classrooms
		  Indoor gyms
			
			20
			60 to 100
			30 to 40
				Street
				  Local
				  Expressway
					
					0.9
					1.4

Garages (auto)
  Parking
  Repair
	
	10
	50
		Stores
		  Stockroom
		  Service area
			
			30
			100
				Car lots
				  Front line
				  Remaining area
					
					100 to 500
					20 to 75

Hospital/Medical
  Lobby
  Dental chair
  Operating table
	
	30
	1000
	2500
		Warehousing, storage
		  Inactive
		  Active
			
			5
			30		

* in fc

Necessary light levels are also affected by the age of the people in that environment. Older people require more light to see as well as younger people. Since recommended light level charts typically assume an average age of 30, they may need to be adjusted if the average age of the people in an environment is other than 30. For an average age of 20, the values can be reduced to 75% of the recommended light levels because younger eyes require less light. However, the recommended light levels are higher for people over the age of 30. The recommended light levels are doubled for an average age of 40, tripled for an average age of 50, and multiplied by four for an average age of 60.

1. Connect the light meter attachment to the correct jacks on the multimeter.

2. Set (draw in) the multimeter function switch for taking light level measurements.

The multimeter and light meter attachment are used to take light level readings in a warehouse parking lot. The readings are 5.6 mV, 6.8 mV, and 6.2 mV for three locations in the parking lot.

3. _____ What is the average light level (in fc)?

4. _____ Is the light level adequate for the application?

5. _____ Is the light meter attachment measuring direct light, indirect light, or both?

6. _____ If an area of the parking lot is used for loading and unloading materials, is the light level adequate for that application?

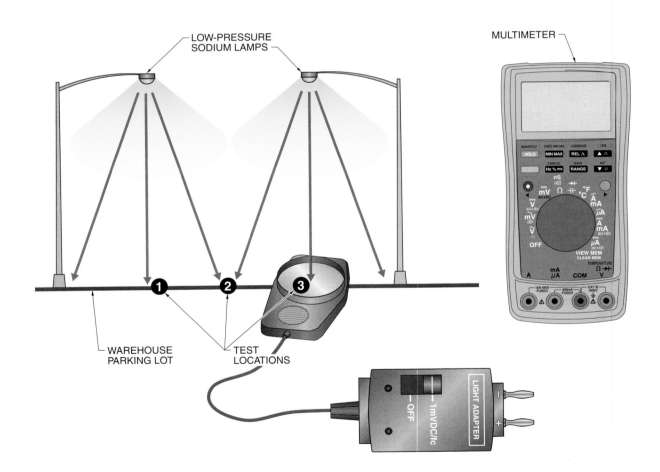

7. Connect the light meter attachment to the correct jacks on the multimeter.

8. Set (draw in) the multimeter function switch for taking light level measurements.

The multimeter and light meter attachment are used to take light level readings inside an active warehouse. The readings are 32.5 mV, 48.6 mV, 50.2 mV, 47.9 mV, 33.3 mV for five locations in the building.

9. _____ What is the average light level (in fc)?

10. _____ Is the light level adequate for the application?

11. _____ If the average age of the warehouse staff is 40, is the light level adequate?

12. _____ Is the light meter attachment measuring direct light, indirect light, or both?

13. Why does the test location in the center of the room have the highest reading?

14. _____ If Lamp 2 burned out, which location would have the highest light level?

15. _____ Which location would have the lowest light level?

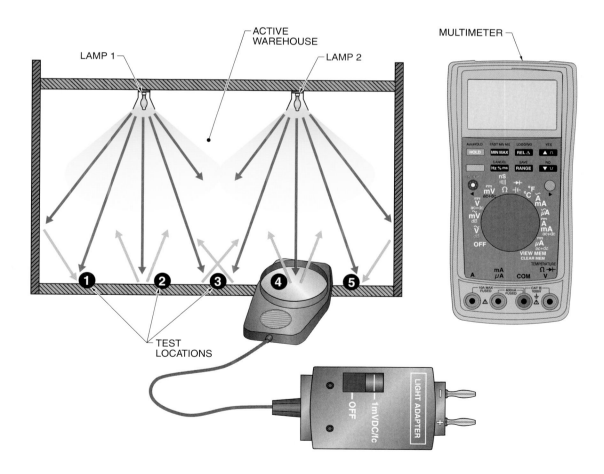

Name_____ Date_____

_____ 1. What are the physical components in a system called?

T   F   2. It is never economical to replace an entire board rather than an individual component.

_____ 3. Which troubleshooting level would include troubleshooting a water pump in a boiler system?

_____ 4. Which troubleshooting method involves using documentation provided by equipment manufacturers?

_____ 5. Which step in the troubleshooting process involves using test instruments extensively to diagnose faulty equipment?

_____ 6. Which troubleshooting level would include troubleshooting an individual winding in an electric motor?

T   F   7. Personnel are expected to document their procedures and findings after troubleshooting equipment or systems.

_____ 8. What part of a system is composed of programs and procedures for operating hardware?

T   F   9. Manuals, wiring diagrams, and other equipment documents are important resources for troubleshooting.

T   F   10. A module is an integrated circuit that is used to create digital circuits.

_____ 11. Which troubleshooting method involves using procedures developed by the facility?

_____ 12. Which troubleshooting level would include troubleshooting a PLC?

T   F   13. Most equipment failures occur during the break-in period.

_____ 14. Which troubleshooting level would include troubleshooting an LCD message display circuit?

_____ 15. What device allows communication between various components used together in a system?

_____ 16. Which type of maintenance is performed to keep equipment and process running with little or no downtime?

_____ 17. Which type of troubleshooting is performed away from the facility with the malfunction?

_____ **18.** Which step in the troubleshooting process involves interviewing the machine operator about malfunctions and modifications?

T   F   **19.** The system recognition step in the troubleshooting process involves researching all the available documentation on a piece of equipment.

_____ **20.** Which step in the troubleshooting process involves determining the cause of the failure?

21. How are equipment failures distributed during typical equipment life expectancy?

22. What are troubleshooter responsibilities in regards to preventing malfunctions on a machine?

23. How could documenting a troubleshooting process help prevent future failures?

24. How are flow charts similar to the six-step troubleshooting procedure?

25. How are the five troubleshooting levels used to diagnose a failure?

Application 10-1
Troubleshooting Two-Way Switching Circuits

A two-way switch is a single-pole, single-throw (SPST) switch used to control a load (such as a lamp) from one location. When in the ON position, the switch makes a connection between two conductors and allows current to flow from one conductor to the other. When in the OFF position, the switch breaks the connection between the two conductors and current cannot flow. Two-way switches have ON and OFF positions clearly marked on them. See Two-Way Switching Diagram.

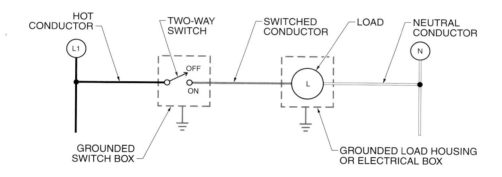

TWO-WAY SWITCHING DIAGRAM

Switches are wired on hot conductors, which are before the load. The conductor between the power supply and the switch remains the "hot" conductor, and the conductor between the switch and the load is called the "switched" conductor. The switched conductor is usually red.

The following tests can be applied to troubleshoot switching circuits. See Switching Circuit Troubleshooting.

1. Test for proper grounding. Connect a voltmeter between the metal box (or the green/bare ground wire in a plastic box) and the hot conductor (ungrounded conductor) going into the switch. There should be system voltage at all times, regardless of the switch position. This test should be done at both the switch box and the load's box. If the system is not properly grounded, ground the system before conducting further tests.

2. Test for correct system voltage. Connect the voltmeter between the hot conductor and the neutral conductor. There should be system voltage at all times, regardless of the switch position.

3. Test the switch for proper operation. Connect the voltmeter between the neutral conductor and switched conductor. There should be no voltage when the switch is in the OFF position and system voltage when the switch is in the ON position.

4. Test the load. Connect the voltmeter between the load terminals. There should be no voltage when the switch is in the OFF position and system voltage when the switch is in the ON position.

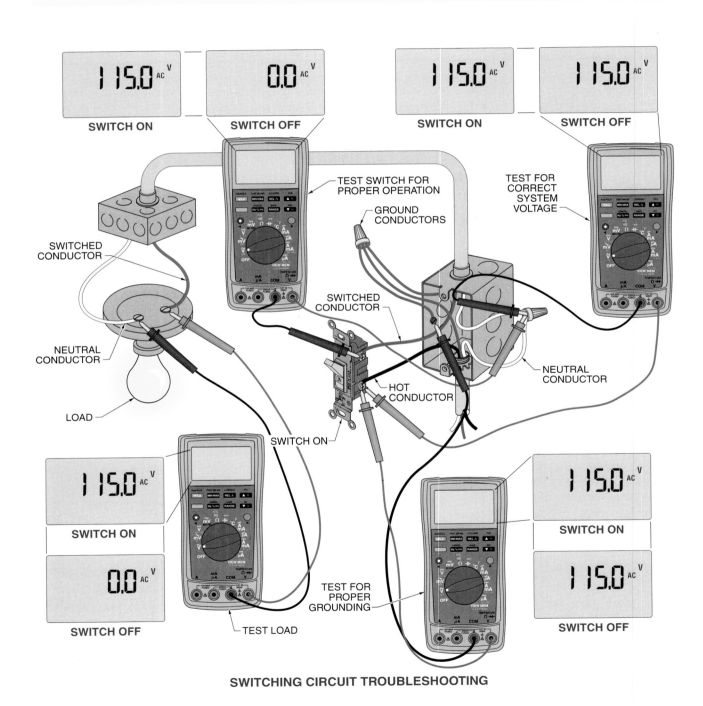

SWITCHING CIRCUIT TROUBLESHOOTING

1. Connect the test leads of Multimeter 1 to test for proper grounding at the box for Switch 1.

2. _____ What should Multimeter 1 read when the switch is in the ON position?

3. _____ What should Multimeter 1 read when the switch is in the OFF position?

4. Connect the test leads of Multimeter 2 to test for correct system voltage out of Switch 1.

5. _____ What should Multimeter 2 read when the switch is in the ON position?

6. _____ What should Multimeter 2 read when the switch is in the OFF position?

7. Connect the test leads of Multimeter 4 to test Switch 2 for proper operation.

8. _____ What should Multimeter 4 read when the switch is in the ON position?

9. _____ What should Multimeter 4 read when the switch is in the OFF position?

10. Connect the test leads of Multimeter 3 to test Lamp 2.

11. _____ What should Multimeter 3 read when the switch is in the ON position?

12. _____ What should Multimeter 3 read when the switch is in the OFF position?

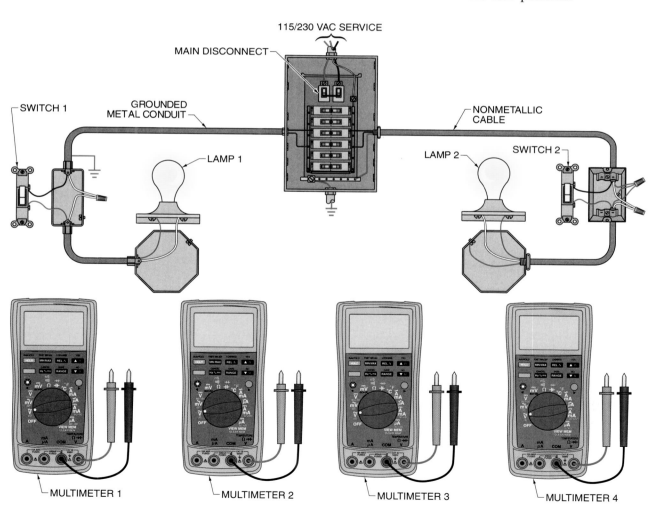

The same tests performed on a two-way switching circuit can be used to troubleshoot a three-way switching circuit. A three-way switch is a single-pole, double-throw (SPDT) switch. Two three-way switches are used to control a load from two different locations. A three-way switch does not have designated ON and OFF positions because the two switches affect each other's operation. The position of one switch that allows the flow of current (turning the load ON) depends on the position of the other switch.

A three-way switch has one common terminal and two traveler terminals. The common terminal is connected to the hot conductor or the switched conductor, depending on the switch's location in the circuit. The traveler terminals are connected to traveler conductors, which connect the two switches. See Three-Way Switching Diagram.

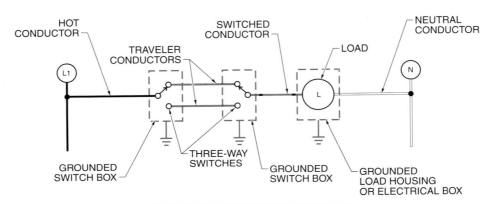

THREE-WAY SWITCHING DIAGRAM

A three-way switch operates like a pair of two-way switches. When the three-way switch is in one position, contact is made with one of the traveler terminals, similar to how a two-way switch makes contact in the ON position. When the three-way switch is in the other position, it makes contact with the other traveler terminal. In either position, one of the travelers is always connected to the common terminal.

When two three-way switches are used together in a circuit, either one can be used to switch the load ON. One switch connects one of the travelers to the circuit. The other switch either completes the circuit (turning the load ON) by connecting its common to the same traveler, or the circuit remains open (the load OFF) by connecting its common to the unconnected traveler.

A three-way switch circuit includes more conductors than a two-way switch, and each conductor must be identified when troubleshooting. The traveler conductors are color-coded red. The switched conductor in a three-way switching circuit is usually red or black and may be marked with colored tape to distinguish it from other conductors. The hot conductor is black and the neutral conductor is white. See Three-Way Switching Circuit.

Power can enter the conduit system at either one of the switches or at the load, but the switches must always be connected before the load. For example, if power enters at the load, the hot conductor is run through the conduit to one of the switches first. This is because it is the hot conductor that must be switched, not the neutral conductor. The travelers and the switched conductor then carry the current back to the load.

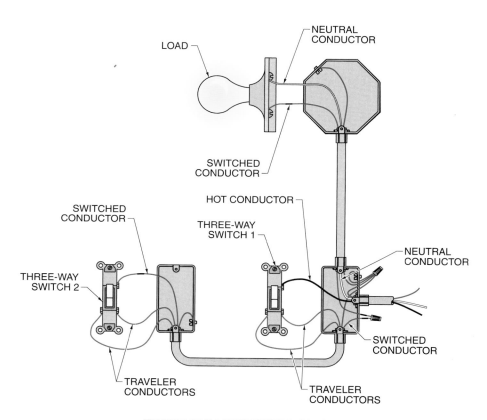

THREE-WAY SWITCHING CIRCUIT

1. _____ What should Multimeter 1 read?

2. _____ Which test is Multimeter 1 performing?

3. _____ If Multimeter 2 reads 115 VAC, what should it read if Switch 2 changes positions?

4. _____ If Multimeter 2 reads 0 VAC, what should it read if Switch 1 changes positions?

5. _____ If Multimeter 3 reads 115 VAC for both positions of Switch 1, is there a problem in Switch 1 or Switch 2?

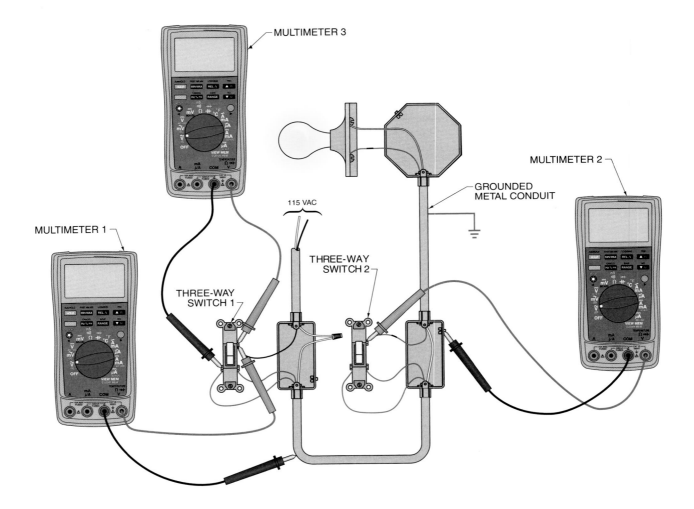

6. _____ If Multimeter 1 reads 115 VAC, what should it read if Switch 1 changes positions?

7. _____ If Multimeter 1 reads 0 VAC, what should it read if Switch 2 changes positions?

8. _____ If Multimeter 1 reads 115 VAC for both positions of Switch 1, is there a problem in Switch 1 or Switch 2?

9. _____ If Multimeter 1 reads 115 VAC for both positions of Switch 2, is there a problem in Switch 1 or Switch 2?

10. _____ What should Multimeter 2 read?

11. _____ If Multimeter 2 reads 0 V, what is the most likely problem?

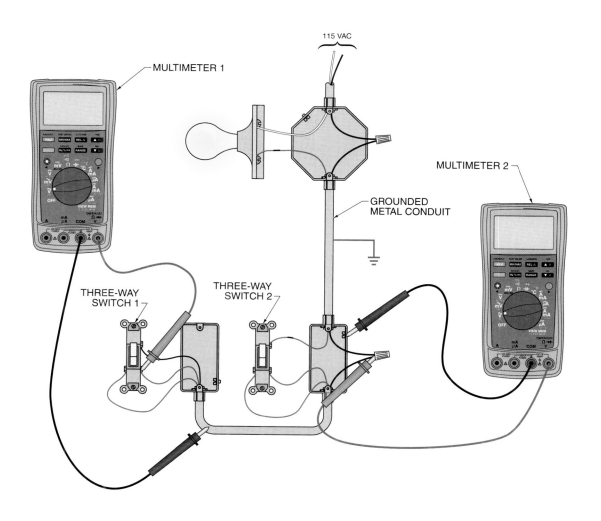

12. _____ If Multimeter 1 reads 115 VAC, what should it read if Switch 1 changes positions?

13. _____ If Multimeter 1 reads 115 VAC, what should it read if Switch 2 changes positions?

14. _____ What should Multimeter 2 read?

15. _____ Will changing the position of Switch 2 change the reading of Multimeter 2?

16. _____ If Multimeter 1 reads 0 VAC for both positions of Switch 1 and Switch 2, could the problem be in the load?

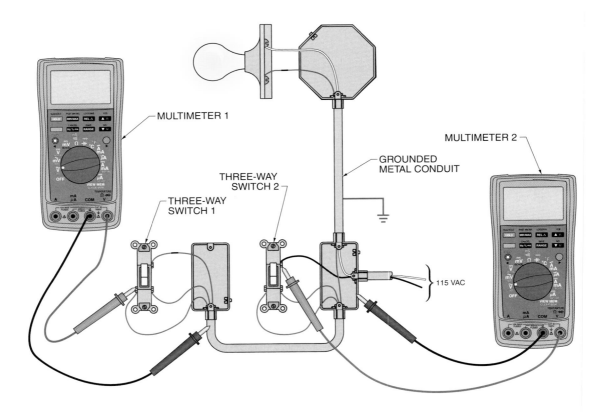

Application 10-3   *Troubleshooting Hardwired Circuits*

Troubleshooting is a systematic process used to find a problem in an electrical circuit. The fact that an electrical circuit can be wired several different ways can complicate the troubleshooting process. A circuit can be hardwired, wired using terminal strips, or wired to a PLC. Each circuit functions exactly the same way, and the wiring type is not obvious to operators, supervisors, or even maintenance personnel until the electrical panels are opened. Each different wiring method requires a different troubleshooting method.

For example, a simple start-stop circuit can be wired by any of the three methods. Although a hardwired control circuit is the most common, it is the most difficult to troubleshoot. See Hardwired Circuit. A control circuit wired to a terminal strip is much easier to troubleshoot because most test instrument measurements can be taken at the terminal strip. See Terminal Strip Wired Circuit. Wiring a simple control circuit using a PLC may seem excessive at first, but with current demands for security and automation, this type of wiring is becoming more common. For example, a PLC can be programmed to automatically start (or stop) the motor for a process, keep track of the number of times the motor is energized, and automatically transmit information about routine maintenance requirements. See PLC-Wired Circuit.

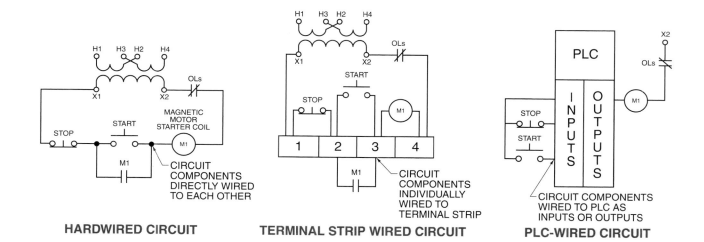

HARDWIRED CIRCUIT   **TERMINAL STRIP WIRED CIRCUIT**   **PLC-WIRED CIRCUIT**

Direct hardwiring is the oldest and most straightforward wiring method used. In direct hardwiring, the power circuit (higher voltage) and control circuit (lower voltage) are wired point-to-point. In point-to-point wiring, each component in a circuit is connected directly to the next component specified on the wiring and line diagram. See Hardwired Reversing Circuit.

For example, the transformer X1 terminal is connected directly to the fuse, the fuse is connected directly to the stop pushbutton, the stop pushbutton is connected directly to the reverse pushbutton, the reverse pushbutton is connected directly to the forward pushbutton, and so on until the final connection from the overload (OL) contact is made back to the X2 terminal.

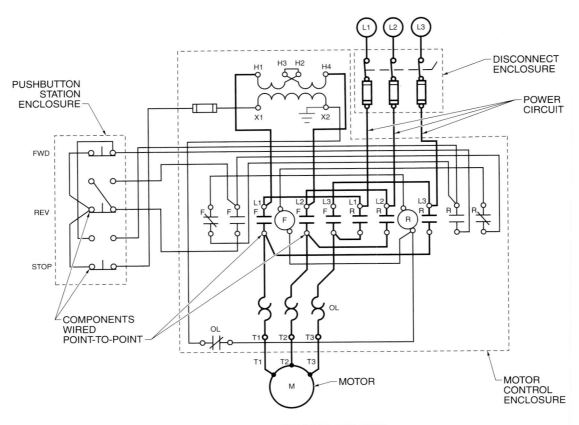

HARDWIRED REVERSING CIRCUIT

The disadvantage of a direct hardwired circuit is that circuit troubleshooting and modification can be time consuming. Gaining access to test points and around conductors with test instruments can be difficult in hardwired circuits. The original wiring may have been intended to be permanent, and test points could be protected by insulation or placed where there may not be enough room to use test leads or current clamps.

When working on hardwired circuits, conductors may need to be manually traced throughout a hardwired circuit, which is time consuming. Wiring diagrams help identify connections and conductor runs, but may not be complete, especially if modifications have been made.

Hardwired circuits are also difficult to modify. Any additional components must be inserted into the circuit without adversely affecting the circuit operation. Determining the necessary connections can be complicated, and the actual wiring may be difficult if conduits and boxes are crowded, if there is not enough room under terminal screws, or if permanent connections (such as soldered connections) must be undone. Modifications also complicate troubleshooting later if they were not documented on the original wiring diagrams.

Some circuit modifications, such as adding indicator lamps, may not cause a problem because they only require adding new conductors, and existing conductors do not need to be moved or removed. Other circuit modifications, such as adding switches, can be more difficult. For example, if forward and reverse limit switches are to be added to a circuit, some of the wiring must first be removed from the circuit, then new wiring for the limit switches must be added. See Modified Hardwired Reversing Circuit.

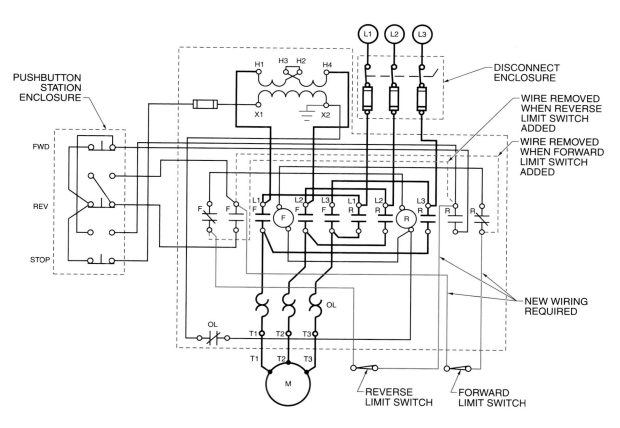

MODIFIED HARDWIRED REVERSING CIRCUIT

The three loads in this circuit are the motor, the forward coil, and the reverse coil. Multimeter 1 takes measurements in the disconnect enclosure after the fuses. Multimeter 3 measures 12.5 A on L1, 13.8 A on L2, and 11.9 A on L3.

1. _____ If Multimeter 1 is connected after Fuse 1 and the motor is running in the forward direction, which loads will contribute to the measured current?

2. _____ If Multimeter 1 is connected after Fuse 2 and the motor is running in the forward direction, which loads will contribute to the measured current?

3. _____ If Multimeter 1 is connected after Fuse 3 and the motor is running in the forward direction, which loads will contribute to the measured current?

4. _____ If Multimeter 2 reads 115 VAC when the reverse pushbutton is pressed, but the coil does not engage, which component (other than the coil itself) is most likely the problem?

5. _____ What is the percentage of current unbalance at the motor?

6. _____ Is the current unbalance acceptable?

7. _____ What should Multimeter 4 read before any pushbutton is pressed?

8. _____ What should Multimeter 4 read when the forward starter coil is on?

9. _____ Will Multimeter 5 read the current to the motor?

10. _____ What prevents both the forward and reverse coils from being energized at the same time?

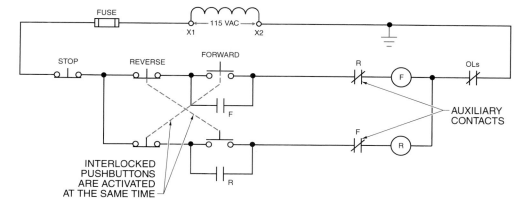

CONTROL CIRCUIT LINE DIAGRAM

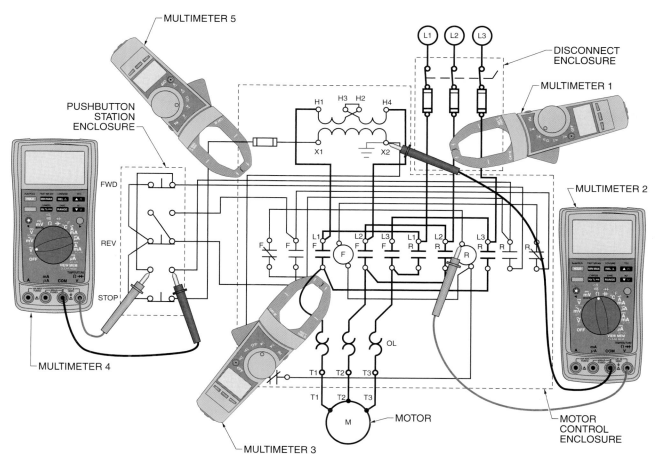

WIRING DIAGRAM

Terminal strip wiring makes circuit troubleshooting quicker and easier because test instrument measurements can be taken directly at the terminal strip. The terminal strip is the connection point for all the components and devices in the circuit. On the line diagram, each component is assigned reference numbers to identify each connection. The wire reference numbers identify the terminals on the terminal strip where the connections are made. For example, the STOP pushbutton is connected between Terminals 1 and 3. The reverse pushbutton has four connections. The normally closed (NC) connections are made between Terminals 3 and 4 and the normally open (NO) connections are made between Terminals 7 and 8. See Line Diagram with Terminal Strip Numbers.

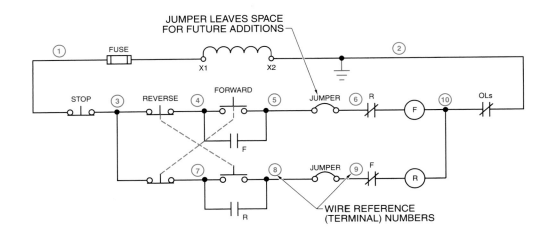

LINE DIAGRAM WITH TERMINAL STRIP NUMBERS

The terminal strip will typically use screw terminals to connect conductors. If many connections are to be made at a certain number, two or more terminals can be assigned to that number to accommodate the conductors. The like-numbered terminals are then connected with jumpers to make them electrically continuous. Jumpers can also be used between two terminal spots with different numbers when future modifications are anticipated. Later, the jumper can be removed and a component added between the two numbers.

Wire reference numbers are assigned from the top left to the bottom right. The power supply connections are usually assigned numbers "1" and "2". When troubleshooting a circuit with a terminal strip, a voltmeter is used to measure the voltage on the power supply terminals to verify the circuit is powered. If the voltage is correct, the black test lead is left on one terminal and the red test lead is moved to different terminals until the problem is located.

Using terminal strips also makes circuit modifications easier because connections are simple to add or undo. Standard options to the circuit, such as indicator lights or additional switches, may be specified with terminal strip numbers already assigned as a wiring guide. See Terminal Strip Wiring Diagram.

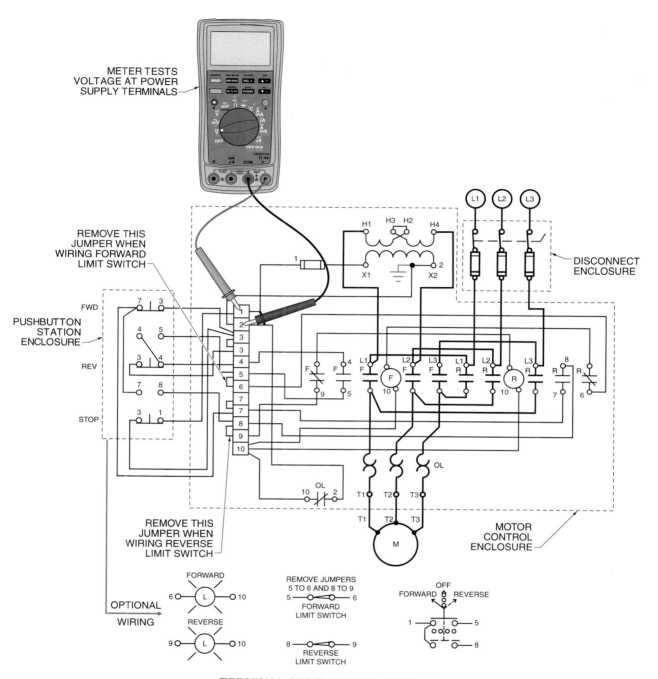

METER TESTS
VOLTAGE AT POWER
SUPPLY TERMINALS

REMOVE THIS
JUMPER WHEN
WIRING FORWARD
LIMIT SWITCH

PUSHBUTTON
STATION
ENCLOSURE

FWD

REV

STOP

DISCONNECT
ENCLOSURE

REMOVE THIS
JUMPER WHEN
WIRING REVERSE
LIMIT SWITCH

MOTOR
CONTROL
ENCLOSURE

OPTIONAL
WIRING

FORWARD

REVERSE

REMOVE JUMPERS
5 TO 6 AND 8 TO 9

5 —o—o— 6
FORWARD
LIMIT SWITCH

8 —o—o— 9
REVERSE
LIMIT SWITCH

OFF
FORWARD   REVERSE

TERMINAL STRIP WIRING DIAGRAM

1. _____ To which terminals is the STOP pushbutton connected?

2. _____ To which terminals is the UP pushbutton connected?

3. _____ To which terminals is the Bottom LS connected?

4. _____ How many components connect to Terminal 6?

5. _____ Which components are connected to Terminal 7?

6. _____ What should Multimeter 1 read when the motor is ON?

7. _____ What should Multimeter 2 read when the motor is OFF?

8. _____ What should Multimeter 3 read when the motor is OFF?

9. _____ What should Multimeter 4 read when the motor is OFF?

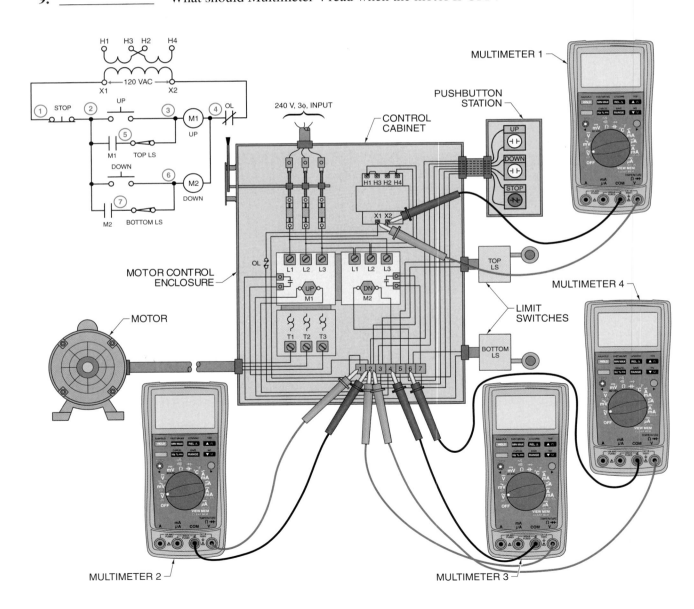

10. _____ What should Multimeter 1 read?

11. _____ If the control circuit fuse is blown (open), what will Multimeter 1 read?

12. _____ What should Multimeter 2 read before any buttons are pressed?

13. _____ What is the problem if Multimeter 3 reads 0 VAC and the fuse is good?

14. _____ What should Multimeter 4 read before any buttons are pressed?

15. _____ If the reverse pushbutton is pressed, what should Multimeter 4 read?

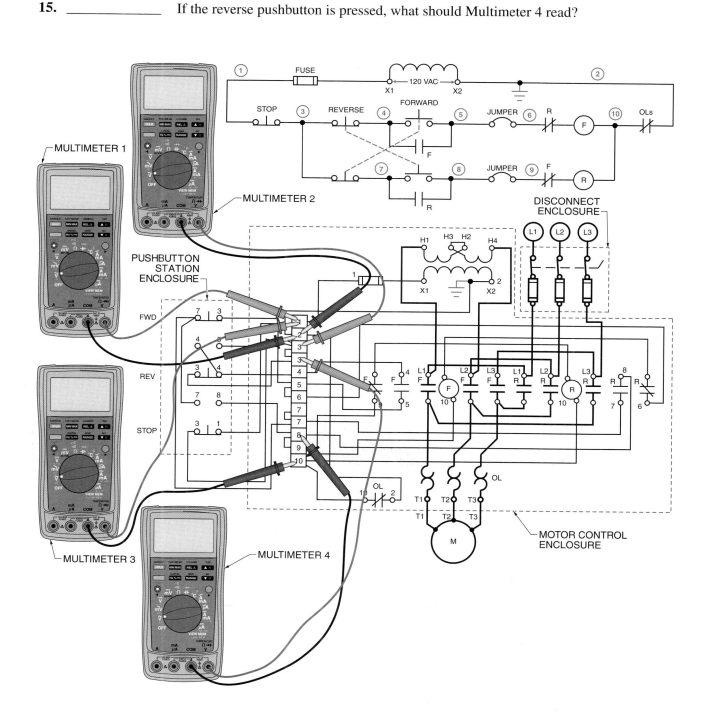

PLCs provide greater flexibility and monitoring capability for control circuits than hardwired or terminal strip circuits. A PLC can monitor all control functions, make decisions based on the programmed logic, and activate loads with digital (ON or OFF) or analog (variable) signals.

A PLC replaces much of a control circuit's wiring. The wiring logic becomes software logic instead, which is much easier to change. The inputs (pushbuttons, limit switches, and overload contacts) are wired directly to the PLC input module and the outputs (motor starter coils and indicator lamps) are wired directly to the PLC output module. The power circuit does not change. See PLC Control Circuit.

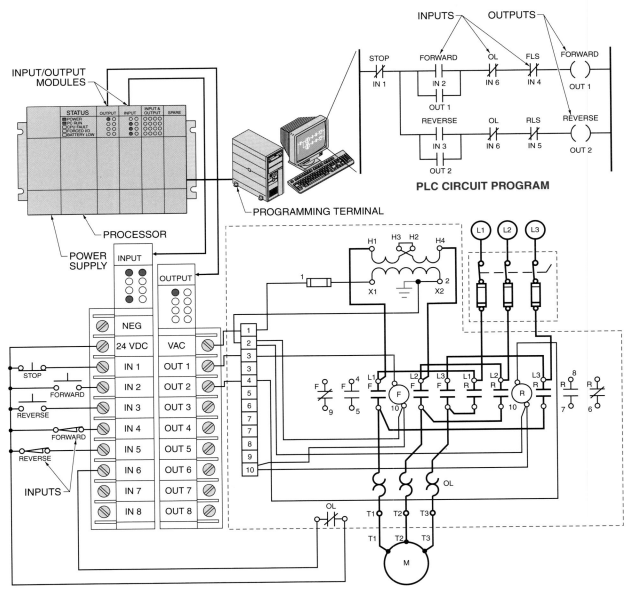

PLC CONTROL CIRCUIT

The circuit operation (logic) is programmed using PLC software and transferred to the PLC. The PLC program monitors and displays the condition (ON or OFF) of the circuit inputs and outputs. If changes in the control circuit are required, they can be reprogrammed and downloaded without changing the circuit wiring.

When programming inputs for a PLC, the actual input type (normally open or normally closed) and the way the input is programmed must be considered. When programming a PLC, a hardwired input can be programmed either normally closed or normally open, even if it is wired only one way. Often, hard-wired normally closed inputs are programmed as normally open inputs. This is because PLCs will only energize outputs preceded by programmed normally closed inputs when the input is activated (open), which is contrary to the wiring logic.

PLCs have several features that are helpful when troubleshooting. Indicator lights on the input and output modules display the current state of the inputs and outputs. If the indicator lights show an error in the control logic (an output is being activated when it should not be), the problem is probably in the PLC program. If the indicator lights show correct logic, but the components are acting unpredictably, the problem is probably in the wiring or component.

Voltage and current can be measured at the input and output modules, which are similar to terminal strips. Also, loads can be forced (activated against the program logic) ON or OFF if necessary. This can be useful for checking the operation of an output without waiting for the PLC to activate the load.

The three loads in this circuit are the motor, the forward coil, and the reverse coil.

1. _____ Which input corresponds to the reverse pushbutton?

2. _____ Which input corresponds to the overloads?

3. _____ Which output corresponds to the forward coil?

4. _____ Which input controls Output 2?

5. _____ Which input affects both outputs?

6. _____ What type and level of voltage is used in the input module?

7. _____ What should Multimeter 1 read when Input 2 is activated?

8. _____ What should Multimeter 2 read before the forward limit switch is activated?

9. _____ Which output will be activated if Input 2 is activated?

10. _____ Which output will be deactivated if Input 5 is activated?

11. _____ If Input 2 is activated, what should Multimeter 3 read?

12. _____ If Input 5 is activated, what should Multimeter 3 read?

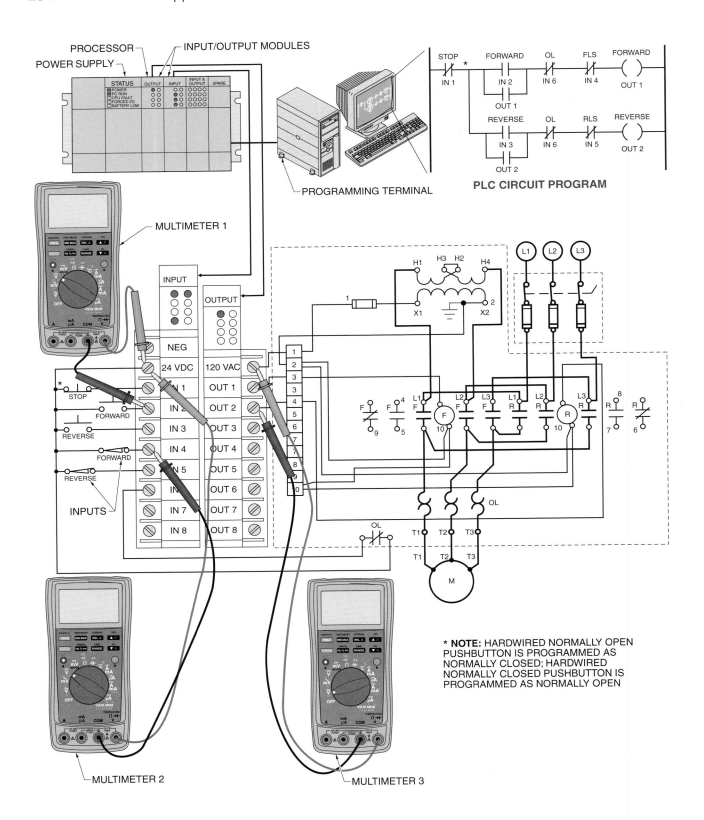

PROCESSOR

POWER SUPPLY

INPUT/OUTPUT MODULES

STATUS
POWER
PC RUN
CPU FAULT
FORCED I/O
BATTERY LOW

OUTPUT   INPUT   INPUT & OUTPUT   SPARE

PROGRAMMING TERMINAL

MULTIMETER 1

INPUT

OUTPUT

NEG
24 VDC   120 VAC
IN 1   OUT 1
IN 2   OUT 2
IN 3   OUT 3
IN 4   OUT 4
IN 5   OUT 5
IN 6   OUT 6
IN 7   OUT 7
IN 8   OUT 8

STOP
FORWARD
REVERSE
FORWARD
REVERSE

INPUTS

MULTIMETER 2

MULTIMETER 3

PLC CIRCUIT PROGRAM

STOP   FORWARD   OL   FLS   FORWARD
IN 1   IN 2   IN 6   IN 4   OUT 1

OUT 1

REVERSE   OL   RLS   REVERSE
IN 3   IN 6   IN 5   OUT 2

OUT 2

* **NOTE:** HARDWIRED NORMALLY OPEN PUSHBUTTON IS PROGRAMMED AS NORMALLY CLOSED; HARDWIRED NORMALLY CLOSED PUSHBUTTON IS PROGRAMMED AS NORMALLY OPEN

METRIC PREFIXES

Multiples and Submultiples	Prefixes	Symbols	Meaning
$1,000,000,000,000 = 10^{12}$	tera	T	trillion
$1,000,000,000 = 10^9$	giga	G	billion
$1,000,000 = 10^6$	mega	M	million
$1000 = 10^3$	kilo	k	thousand
$100 = 10^2$	hecto	h	hundred
$10 = 10^1$	deka	da	ten
Unit $1 = 10^0$			
$.1 = 10^{-1}$	deci	d	tenth
$.01 = 10^{-2}$	centi	c	hundredth
$.001 = 10^{-3}$	milli	m	thousandth
$.000001 = 10^{-6}$	micro	μ	millionth
$.000000001 = 10^{-9}$	nano	n	billionth
$.000000000001 = 10^{-12}$	pico	p	trillionth

METRIC CONVERSIONS

Initial Units	Final Units											
	giga	mega	kilo	hecto	deka	base unit	deci	centi	milli	micro	nano	pico
giga		3R	6R	7R	8R	9R	10R	11R	12R	15R	18R	21R
mega	3L		3R	4R	5R	6R	7R	8R	9R	12R	15R	18R
kilo	6L	3L		1R	2R	3R	4R	5R	6R	9R	12R	15R
hecto	7L	4L	1L		1R	2R	3R	4R	5R	8R	11R	14R
deka	8L	5L	2L	1L		1R	2R	3R	4R	7R	10R	13R
base unit	9L	6L	3L	2L	1L		1R	2R	3R	6R	9R	12R
deci	10L	7L	4L	3L	2L	1L		1R	2R	5R	8R	11R
centi	11L	8L	5L	4L	3L	2L	1L		1R	4R	7R	10R
milli	12L	9L	6L	5L	4L	3L	2L	1L		3R	6R	9R
micro	15L	12L	9L	8L	7L	6L	5L	4L	3L		3R	6R
nano	18L	15L	12L	11L	10L	9L	8L	7L	6L	3L		3R
pico	21L	18L	15L	14L	13L	12L	11L	10L	9L	6L	3L	

OHM'S LAW AND POWER FORMULA

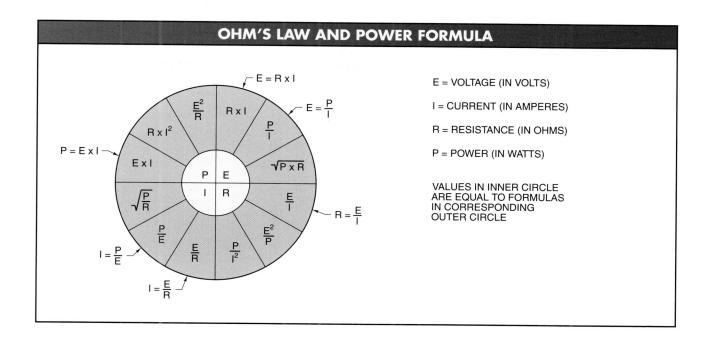

E = VOLTAGE (IN VOLTS)

I = CURRENT (IN AMPERES)

R = RESISTANCE (IN OHMS)

P = POWER (IN WATTS)

VALUES IN INNER CIRCLE
ARE EQUAL TO FORMULAS
IN CORRESPONDING
OUTER CIRCLE

VOLTAGE CONVERSIONS

To Convert	To	Multiply By
rms	Average	0.9
rms	Peak	1.414
Average	rms	1.111
Average	Peak	1.567
Peak	rms	0.707
Peak	Average	0.637
Peak	Peak-to-peak	2

POWER CONVERSIONS

Power	W	ft-lb/s	HP	kW
Watt	1	0.7376	0.000341	0.001
Foot-pound/sec	1.356	1	0.000818	0.001356
Horsepower	745.7	550	1	0.7457
Kilowatt	1000	736.6	1.341	1

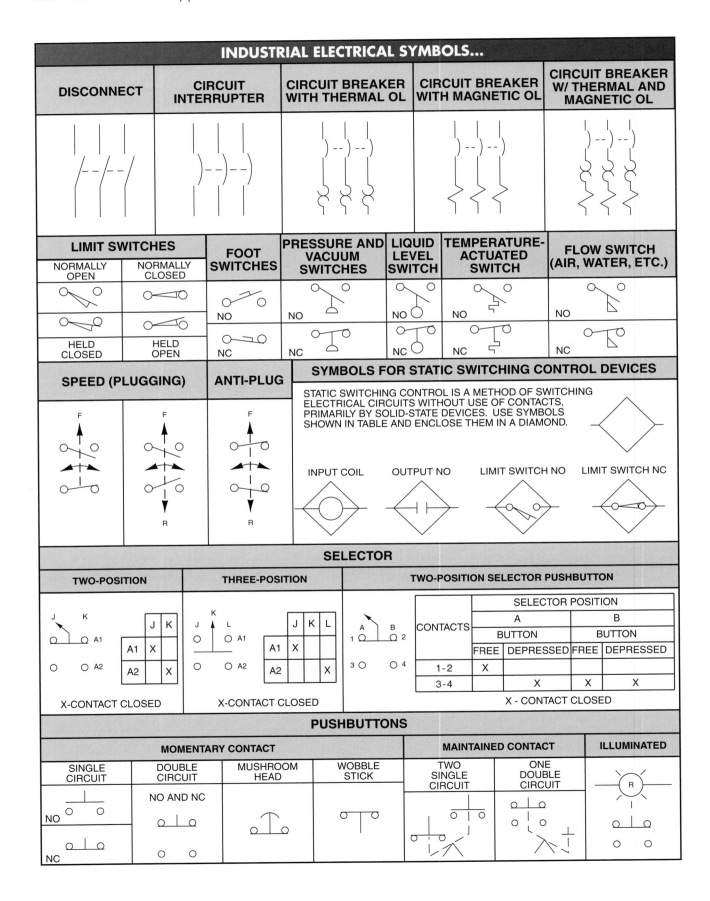

INDUSTRIAL ELECTRICAL SYMBOLS...

| DISCONNECT | CIRCUIT INTERRUPTER | CIRCUIT BREAKER WITH THERMAL OL | CIRCUIT BREAKER WITH MAGNETIC OL | CIRCUIT BREAKER W/ THERMAL AND MAGNETIC OL |

LIMIT SWITCHES / FOOT SWITCHES / PRESSURE AND VACUUM SWITCHES / LIQUID LEVEL SWITCH / TEMPERATURE-ACTUATED SWITCH / FLOW SWITCH (AIR, WATER, ETC.)

LIMIT SWITCHES		FOOT SWITCHES	PRESSURE AND VACUUM SWITCHES	LIQUID LEVEL SWITCH	TEMPERATURE-ACTUATED SWITCH	FLOW SWITCH (AIR, WATER, ETC.)
NORMALLY OPEN	NORMALLY CLOSED	NO	NO	NO	NO	NO
HELD CLOSED	HELD OPEN	NC	NC	NC	NC	NC

SPEED (PLUGGING) / ANTI-PLUG / SYMBOLS FOR STATIC SWITCHING CONTROL DEVICES

STATIC SWITCHING CONTROL IS A METHOD OF SWITCHING ELECTRICAL CIRCUITS WITHOUT USE OF CONTACTS, PRIMARILY BY SOLID-STATE DEVICES. USE SYMBOLS SHOWN IN TABLE AND ENCLOSE THEM IN A DIAMOND.

INPUT COIL OUTPUT NO LIMIT SWITCH NO LIMIT SWITCH NC

SELECTOR

TWO-POSITION

	J	K
A1	X	
A2		X

X-CONTACT CLOSED

THREE-POSITION

	J	K	L
A1	X		
A2			X

X-CONTACT CLOSED

TWO-POSITION SELECTOR PUSHBUTTON

CONTACTS	SELECTOR POSITION			
	A		B	
	BUTTON		BUTTON	
	FREE	DEPRESSED	FREE	DEPRESSED
1-2	X			
3-4		X	X	X

X - CONTACT CLOSED

PUSHBUTTONS

MOMENTARY CONTACT				MAINTAINED CONTACT		ILLUMINATED
SINGLE CIRCUIT	DOUBLE CIRCUIT	MUSHROOM HEAD	WOBBLE STICK	TWO SINGLE CIRCUIT	ONE DOUBLE CIRCUIT	
NO / NC	NO AND NC					

...INDUSTRIAL ELECTRICAL SYMBOLS...

CONTACTS

INSTANT OPERATING				TIMED CONTACTS - CONTACT ACTION RETARDED AFTER COIL IS:			
WITH BLOWOUT		WITHOUT BLOWOUT		ENERGIZED		DE-ENERGIZED	
NO	NC	NO	NC	NOTC	NCTO	NOTO	NCTC

OVERLOAD RELAYS

THERMAL	MAGNETIC

SUPPLEMENTARY CONTACT SYMBOLS

SPST NO		SPST NC		SPDT		TERMS
SINGLE BREAK	DOUBLE BREAK	SINGLE BREAK	DOUBLE BREAK	SINGLE BREAK	DOUBLE BREAK	SPST SINGLE-POLE, SINGLE-THROW SPDT SINGLE-POLE, DOUBLE-THROW DPST DOUBLE-POLE, SINGLE-THROW DPDT DOUBLE-POLE, DOUBLE-THROW NO NORMALLY OPEN NC NORMALLY CLOSED

DPST, 2NO		DPST, 2NC		DPDT	
SINGLE BREAK	DOUBLE BREAK	SINGLE BREAK	DOUBLE BREAK	SINGLE BREAK	DOUBLE BREAK

METER (INSTRUMENT)

INDICATE TYPE BY LETTER	TO INDICATE FUNCTION OF METER OR INSTRUMENT, PLACE SPECIFIED LETTER OR LETTERS WITHIN SYMBOL.			
	AM or A	AMMETER	VA	VOLTMETER
	AH	AMPERE HOUR	VAR	VARMETER
	µA	MICROAMMETER	VARH	VARHOUR METER
	mA	MILLAMMETER	W	WATTMETER
	PF	POWER FACTOR	WH	WATTHOUR METER
	V	VOLTMETER		

PILOT LIGHTS

INDICATE COLOR BY LETTER	
NON PUSH-TO-TEST	PUSH-TO-TEST

INDUCTORS

IRON CORE
AIR CORE

COILS

		DUAL-VOLTAGE MAGNET COILS		BLOWOUT COIL
		HIGH-VOLTAGE	LOW-VOLTAGE	
		LINK	LINKS	
		1 2 3 4	1 2 3 4	

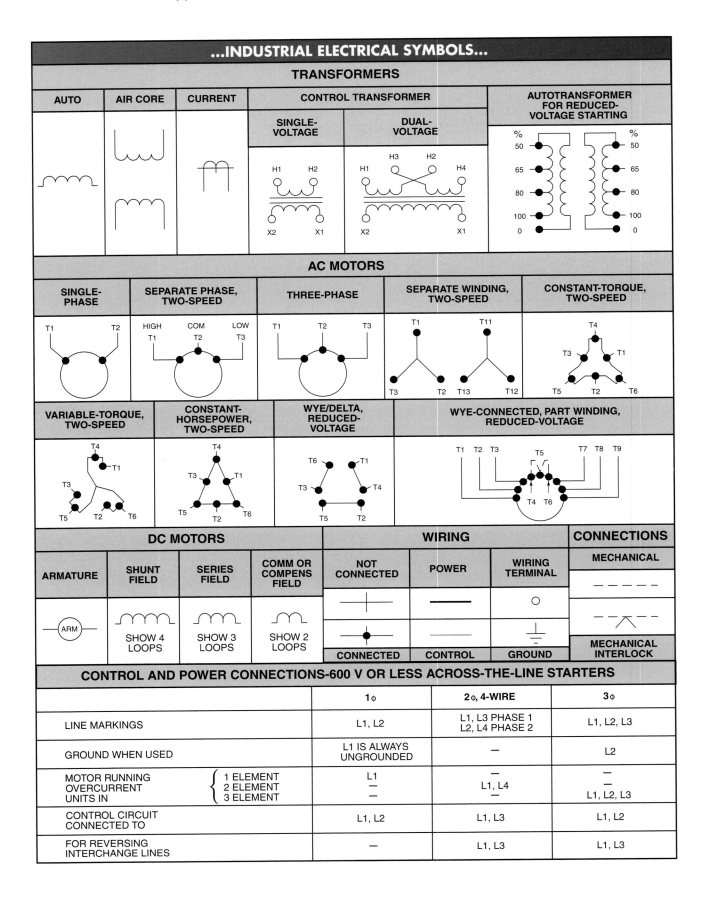

...INDUSTRIAL ELECTRICAL SYMBOLS...

TRANSFORMERS

AUTO	AIR CORE	CURRENT	CONTROL TRANSFORMER		AUTOTRANSFORMER FOR REDUCED-VOLTAGE STARTING
			SINGLE-VOLTAGE	DUAL-VOLTAGE	

AC MOTORS

SINGLE-PHASE	SEPARATE PHASE, TWO-SPEED	THREE-PHASE	SEPARATE WINDING, TWO-SPEED	CONSTANT-TORQUE, TWO-SPEED

VARIABLE-TORQUE, TWO-SPEED	CONSTANT-HORSEPOWER, TWO-SPEED	WYE/DELTA, REDUCED-VOLTAGE	WYE-CONNECTED, PART WINDING, REDUCED-VOLTAGE

DC MOTORS / WIRING / CONNECTIONS

DC MOTORS				WIRING			CONNECTIONS
ARMATURE	SHUNT FIELD	SERIES FIELD	COMM OR COMPENS FIELD	NOT CONNECTED	POWER	WIRING TERMINAL	MECHANICAL
ARM	SHOW 4 LOOPS	SHOW 3 LOOPS	SHOW 2 LOOPS	CONNECTED	CONTROL	GROUND	MECHANICAL INTERLOCK

CONTROL AND POWER CONNECTIONS-600 V OR LESS ACROSS-THE-LINE STARTERS

		1φ	2φ, 4-WIRE	3φ
LINE MARKINGS		L1, L2	L1, L3 PHASE 1 L2, L4 PHASE 2	L1, L2, L3
GROUND WHEN USED		L1 IS ALWAYS UNGROUNDED	—	L2
MOTOR RUNNING OVERCURRENT UNITS IN	1 ELEMENT	L1	—	—
	2 ELEMENT	—	L1, L4	—
	3 ELEMENT	—	—	L1, L2, L3
CONTROL CIRCUIT CONNECTED TO		L1, L2	L1, L3	L1, L2
FOR REVERSING INTERCHANGE LINES		—	L1, L3	L1, L3

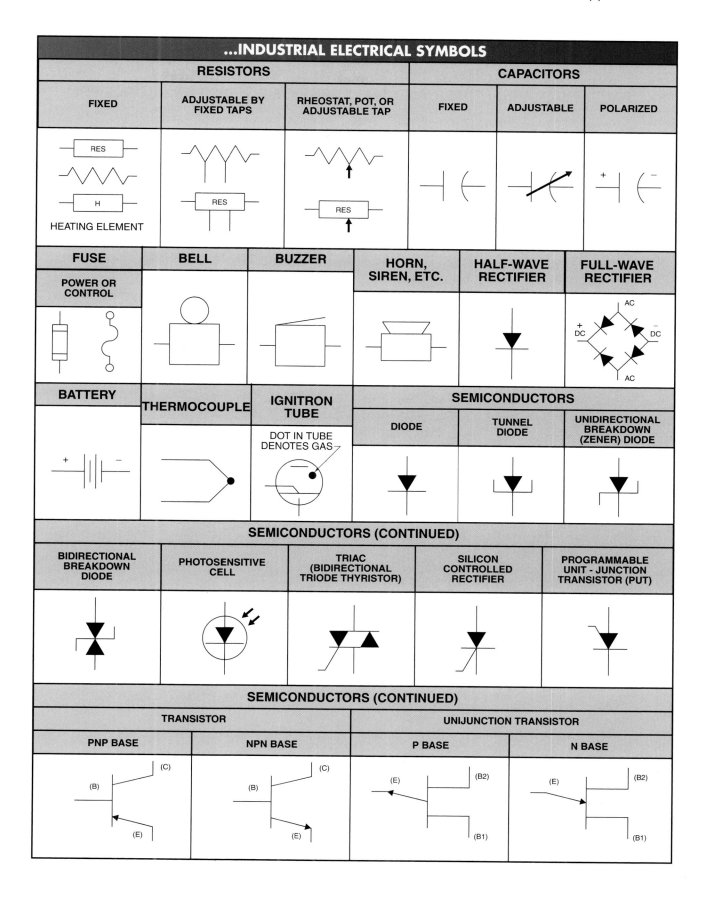

POWER QUALITY TROUBLESHOOTING CHECKLIST
FACILITY TRANSFORMER AND MAIN SERVICE EQUIPMENT PROBLEMS

Problem Observed or Reported:
❏ CBs Tripping/Fuses Blowing ❏ Conduit Overheating ❏ Overheated Neutrals ❏ Electrical Shocks
❏ Damaged Equipment ❏ Other _____

Distribution Type:
❏ 1φ ❏ 3φY ❏ 3φΔ ❏ Fuses ❏ CBs ❏ Voltage(s) _____V ❏ Amperage Rating _____ A
❏ Other _____

Problem Pattern:
Day(s) of week:
❏ Continuous ❏ Random ❏ Monday ❏ Tuesday ❏ Wednesday ❏ Thursday ❏ Friday ❏ Saturday ❏ Sunday
Time(s):
❏ Continuous ❏ Random ❏ Always Same Time _____ ❏ Morning ❏ Afternoon ❏ Evening ❏ Night

Problem or Distribution History:
Has problem been observed or reported before? ❏ No ❏ Yes _____
Was any corrective action taken? ❏ No ❏ Yes _____
Are other parts of system affected? ❏ No ❏ Yes _____
Have additional loads been added to system? ❏ No ❏ Yes _____
Has there been any recent work or changes made to system lately? ❏ No ❏ Yes _____
Are large power loads being switched ON/OFF? ❏ No ❏ Yes _____
Is main service panel properly grounded? ❏ No ❏ Yes _____
Are any subpanels grounded? ❏ No ❏ Yes _____
Has there been a recent lightning storm? ❏ No ❏ Yes _____
Has there been a recent utility feeder outage? ❏ No ❏ Yes_____

Possible Problem(s):
❏ Conductors Undersized (Hot) ❏ Neutral Conductors Shared/Undersized ❏ High Number of Nonlinear Loads
❏ Voltage/Current Unbalance ❏ Harmonics ❏ System Undersized ❏ Improper Wiring/Grounding
❏ Other _____

Measurements Taken:
Taken at Panel _____ Located at _____
Voltage ____V, Current _____ A, Power _____W, _____VA, _____VAR, Power Factor _____ PF, _____ DPF
Voltage THD _____, Current THD _____, K Factor _____, Other _____

Waveform Shape:
Voltage Waveform:
❏ Sinusoidal ❏ Non-Sinusoidal ❏ Flat-Topped ❏ Other _____
Current Waveform:
❏ Sinusoidal ❏ Non-Sinusoidal ❏ Pulsed ❏ Other _____

Measurements Taken at Load (Over Time):
Normal Voltage _____V, Lowest Sag _____V, Highest Swell _____ V
Number of transients recorded _____ over a time period of _____ at a level of _____% above normal

Possible Problem Solution(s):
❏ Oversize Neutrals ❏ Run Separate Neutrals ❏ Additional Transformer ❏ Harmonic Filter ❏ Change to K-Rated Transformer
❏ Add Subpanel ❏ Separate Loads ❏ Proper Wiring and Grounding ❏ Proper Fuses/CBs/Monitors
❏ Power Factor Correction Capacitors ❏ Change Transformer Size ❏ Surge Suppressor
❏ Other _____

DIGITAL MULTIMETER

DIGITAL MULTIMETER

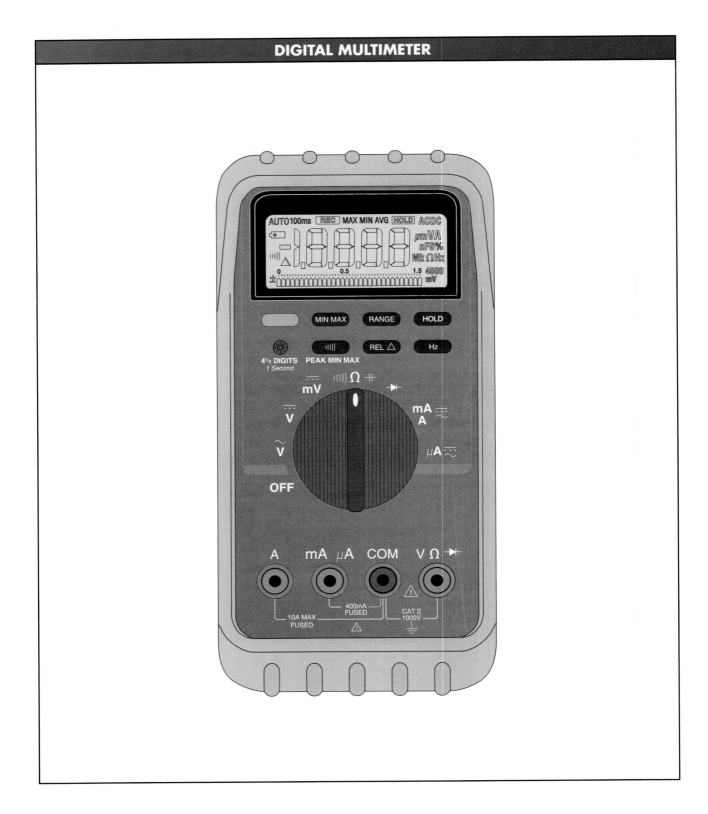

CLAMP METER

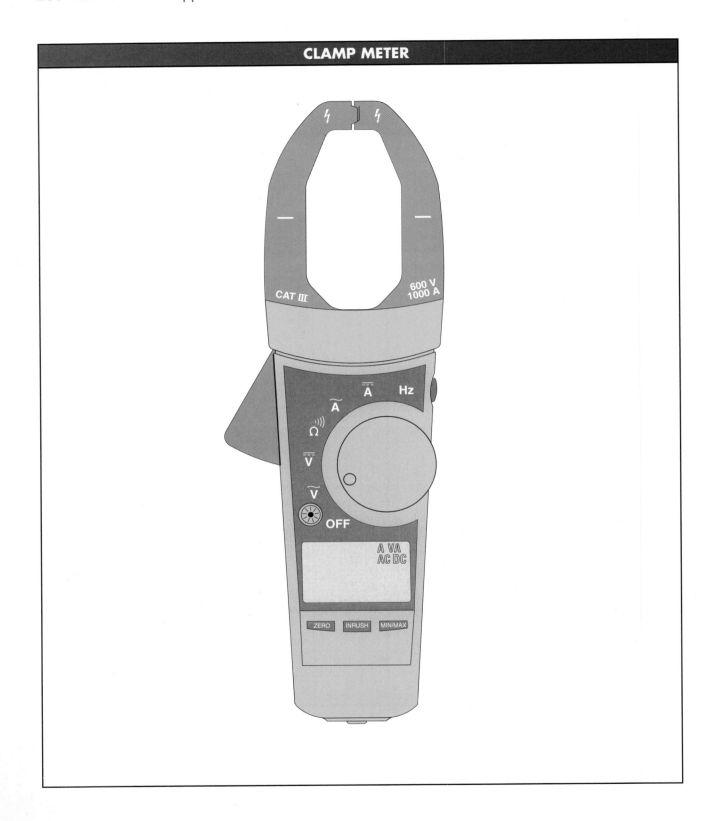